A R McNicoll

PROTEIN INTERACTIONS

PROTEIN INTERACTIONS

Gregorio Weber

Chapman and Hall
New York   London

First published in 1992 by
Chapman and Hall
an imprint of
Routledge, Chapman & Hall, Inc.
29 West 35 Street
New York, NY 10001

Published in Great Britain by

Chapman and Hall
2-6 Boundary Row
London SE1 8HN

Library of Congress Cataloging in Publication Data

Weber, Gregorio.
    Protein interactions/Gregorio Weber.
        p.   cm.
    Includes bibliographical references and index.
    ISBN 0-412-03031-4
    1. Protein binding.   I. Title.
QP551.W38   1992   91-28667
574.19′245—dc20   CIP

British Library Cataloging in Publication Data

Weber, Gregorio
    Protein interactions.
    I. Title
    547.75

    ISBN 0412030314

Dedicated to Those Who Put Doubt Above Belief

Contents

Preface

This book contains a discussion of concepts that facilitate the understanding of the simpler aspects of the interactions of proteins with physiological ligands and with other proteins.

I attempt here generalizations that often demand mathematical derivation and numerical computation, which I prefer to present graphically. In these days of naive confidence in computer calculations it seems necessary to restate that no computation can be more reliable than the concepts that underlie it, and that the most useful lesson that we have learned from computers is that the grammar is more important than the numbers.

At any time and in any scientific subject it is comparatively easy to master the concepts that govern what is already understood and widely practiced. It is far more difficult to appreciate that simple concepts, although demonstrably valid in known cases, cannot be extended to all systems regardless of complexity. Physical chemistry is an area of science greatly burdened by overconfidence in the universal value of simple rules, and in applying these to the proteins I have tried to make a clear distinction between what we can and cannot take for granted.

In my exposition I often stress the limits of our present knowledge of proteins on two counts. First, because awareness of present deficiencies is the origin of future knowledge. Second, because appreciation of our present limitations in the area of protein interactions provides a salutary antidote to the impression of perfection that the student of biochemistry is likely to receive from the amount of structural detail of the proteins that X-ray crystallography, and more recently magnetic resonance, have made available.

I have lived through the period of rapid expansion of the experimental techniques of protein investigation and have thus witnessed the first steps of many subjects that are now presented to the student without reference to their beginnings. When possible I have tried to refer to the original research rather than to recent reviews of these subjects. For it is not difficult to notice that the first observations contain in their naive exposition of facts and concepts much more than many subsequent elaborations.

Urbana, July 1991

I

Thermodynamic Fundamentals

Energy and Entropy

In discussions of energetics the word system denotes an isolated portion of matter of arbitrary composition. The energy, or more precisely the internal energy U of a system is that characteristic property that is increased by the absorption of heat, Q (measured in calories) and decreased by the performance of external work w (measured in ergs or Joules) [1]. The first law of thermodynamics states that the changes in internal energy dU are related to the changes in heat, dQ and work, dw in the manner

$$dU = dQ - dw \qquad (1)$$

Any series of events that take the system out of its initial condition and finally brings it back to its original state perform a cycle in which $dU = 0$. We shall be exclusively concerned with system transformations, cyclic or otherwise, carried out at constant pressure p, whether this be the atmospheric pressure or a different one. For such cases

$$dw = p\,dV \qquad (2)$$

where dV is the change in volume of the system. Clearly dw represents the work incurred in the expansion or contraction of the system against the constant pressure. It follows that

$$dU = dQ - p\,dV \qquad (3)$$

If the internal energy remains unchanged ($dU = 0$)

$$dQ = p\,dV \qquad (4)$$

an equation that postulates the equivalence of heat and mechanical work. The numerical equivalence is 1 calorie = 4.155 J or 4.155×10^7 ergs. It is convenient to express the internal energy of the system by means of the enthalpy H, a quantity somewhat different from U. A system of volume V at pressure p has an enthalpy

$$H = U + pV \tag{5}$$

A change in enthalpy entails

$$dH = dU + p\,dV + V\,dp \tag{6}$$

and from (3)

$$dH = dQ + V\,dp \tag{7}$$

The last equation indicates that the change in enthalpy equals the change in heat content at constant pressure.

The second law of thermodynamics starts with the definition of the entropy: If a system changes from an initial to a final state its change in entropy dS is given by

$$dS = S_{\text{final}} - S_{\text{initial}} = \sum dQ_i/T_i \tag{8}$$

The entropy change is therefore the sum, with appropriate sign, of the changes in heat content taking place at temperature T divided by the temperature at which they take place. The number of terms in the sum of Eq. (8) must be large enough so that T_i is a virtual constant during the absorption of an amount of heat dQ_i. It was first deduced by Clausius that while the state of a system does not imply a unique value for the heat content it does define a unique entropy. Heat can be converted into work when a quantity of heat is transferred from a higher temperature, T_h to a lower one T_1. In a cycle of heat transfer operations that returns the system to the original conditions (Carnot cycle) the yield of work q is given by

$$q = \frac{\text{work performed}}{\text{heat transferred}} \leq 1 - (T_1/T_h) \tag{9}$$

The performance depends therefore on the absolute temperatures of the source (T_h) and sink (T_1) of heat and is always conspicuously less than 1. The second law of thermodynamics generalizes the incomplete transformation of heat into work by stating that in a spontaneous process (isolated system) external work cannot be performed at the expense of a decrease in the entropy of the system. Accordingly in Eq. (8)

$$dS(\text{isolated system}) > 0 \qquad (10)$$

In all spontaneous processes occurring in an isolated system the entropy must necessarily increase.

The Thermodynamic Potentials

The quantities U and H are uniquely determined by the variables of state p, T, V, and S the latter being determined by the previous three variables and the *specific composition of the system*. U and H are potential functions in the sense that the difference between the initial and final values determines the capacity of the system to perform work when it passes from one state to another. To these thermodynamic potentials a third, the most important one for the chemist, must be added [2]. This is the free energy, or more specifically the Gibbs free energy G, formally defined by the relation

$$G = H - TS \qquad (11)$$

Small changes in G are related to those in H, T, and S by

$$dG = dH - T\,dS - S\,dT \qquad (12)$$

which on introduction of Eqs. (7) and (8) becomes

$$dG = V\,dp - S\,dT \qquad (13)$$

It is seen in Table 1 that the potentials are determined by the values of V, p, T, and S. p and T are external variables in the sense that they do not belong to the system itself and V defines an extensive property that can be varied without changing the relative proportions of the components. Thus the properties of the thermodynamic potentials that derive from the nature of the components of the system and their relative amounts are contained in the unique value of the entropy. Equation (12) gives one of the most important properties of the free energy: At constant temperature and pressure G cannot change value unless S changes. In a spontaneous process $dS > 0$. Therefore if such spontaneous process occurs at constant temperature and pressure $dG < 0$: the free energy of the system necessarily decreases. Any external work performed by the system under this condition must be done at the expense of its free energy. The importance of the thermodynamic potential G becomes evident when we

Table 1. Summary of Thermodynamic Potentials

Symbol	Independent variables	Important relations	Name
U	V, S	$T = dU/dS$ $dU = T\,dS - p\,dV$ $p = -(dU/dV)_S$	Energy
H	p, S	$T = dH/dS$ $dH = T\,dS + V\,dp$ $V = (dH/dp)_S$	Enthalpy
G	p, T	$V = dG/dp$ $dG = -S\,dT + V\,dp$ $S = -(dG/dT)_p$	Free energy

consider the relations:

$$dV = \left(\frac{dG}{dp}\right)_T ; \qquad dS = -\left(\frac{dG}{dT}\right)_p \qquad (14)$$

which permit the determination of the changes in volume and entropy from the changes in the Gibbs free energy with pressure, at constant temperature and with temperature, at constant pressure, respectively. A further important property of the free energy is contained in the Gibbs–Helmholz relation:

$$d\left(\frac{dG}{T}\right)\Big/d\left(\frac{1}{T}\right) = dH \qquad (15)$$

This relation permits one to determine the enthalpy change of the system from the thermal coefficient of the free energy change. It is derived by noting that from (11) and (14),

$$G = H + T\left(\frac{dG}{dT}\right); \qquad \frac{dG}{dT} = \frac{-dG/d(1/T)}{T^2}$$

which yield

$$\frac{G + [dG/d(1/T)]}{T} = \frac{d(G/T)}{1/T} = H \qquad (16)$$

Replacing G and H by their finite increments dG and dH gives Eq. (15). Relations (14) and (15) demonstrate that the changes in volume, entropy, and enthalpy occurring with a specified change of the system, for example, when a chemical reaction takes place, can be found through the changes in free energy with temperature and pressure. In the relation $G = H - TS$ only two of the three quantities G, H, and S are independent. These may be taken to be H and S, the enthalpy and entropy, respectively. Their relative importance, particularly as regards chemical reactions, may be conveyed by paraphrasing the words of Emden [3]: "in the factory of nature entropy has the position of manager, for it dictates the manner and method of the whole business, while energy merely does the book-keeping balancing credits and debts."

Gibbs Free Energy and the Chemical Potential

In a homogeneous system made up of a single component a change in Gibbs free energy cannot take place, according to Eq. (13), unless $dp <> 0$ or $dT <> 0$. Such a homogeneous system could well be called a "physical system." When several distinct components are present it is necessary to take into account the possibility of chemical reactions that can change their amounts, and therefore the value of S, although the system is isolated and remains at constant temperature and pressure. Let such "chemical system" involve components A, B, C, D capable of undergoing chemical reaction. Then, if in Eq. (13) $dT = dp = 0$, we have instead of $dG = 0$

$$dG = \mu(A)\,dn(A) + \mu(B)dn(B) + \mu(C)\,dn(C) + \mu(D)\,dn(D) \quad (17)$$

In this equation $dn(A)\ldots dn(D)$ are the molar amounts by which the components change in the chemical reaction. $\mu(A)\ldots\mu(D)$ represent the changes in Gibbs free energy that take place in a change of 1 mol of the substance, under the conditions of composition, temperature, and pressure peculiar to the system. Equation (17) may then be considered as providing a definition of $\mu(A)\ldots\mu(D)$, the "chemical potentials." At constant pressure and temperature the whole of the change in the free energy derives from the changes in the chemical potentials.

The biochemist—always—and the chemist—more often than not—are concerned with reactions taking place in solution. It is empirically known that chemical reactions in solution proceed to an extent determined by the concentrations of the reagents, a relation expressed in the so-called law of mass action [4]. Therefore the changes in the chemical potentials of the reagents must depend on their concentration. This dependence is taken into account by writing the chemical potential in the form

$$\mu = \mu_0 + w(c, c_0) \tag{18}$$

According to (18) the chemical potential results from two contributions: μ_0, the standard chemical potential, designates the part of the potential that can be known only through the chemical reactions in which the substance takes part when present at an arbitrarily fixed concentration c_0. This is commonly taken to be 1 mol/liter. The concentration dependent function $w(c, c_0)$ corresponds to the free energy change required to bring this 1 mol of reagent from the standards concentration c_0 to its actual concentration c. It is calculated from the dependence of the osmotic pressure on the concentration [5]. If the dissolved molecules behaved in an ideal fashion they would exert on a semipermeable membrane separating them from pure solvent an osmotic pressure π dependent on the concentration in the same manner as the pressure exerted on the walls of the container by the molecules of an ideal gas. The pressure developed at a molar concentration c would then equal

$$\pi = cRT \tag{19}$$

where R is the familiar gas constant. The free energy change in the dilution from c to c_0 equals the osmotic work

$$w = \int_c^{c_0} \pi \, dV \tag{20}$$

A solution of concentration c supposes a molar volume $V = 1/c$. Therefore introducing the relation $dV = -dc/c^2$ and Eq. (19), Eq. (20) becomes

$$w = \int_c^{c_0} \frac{-RT \, dc}{c} = RT \ln\left(\frac{c}{c_0}\right) \tag{21}$$

and for $c_0 = 1$

$$\mu = \mu_0 + RT \ln c \tag{22}$$

Real solutions do not follow this simple equation except at low concentrations. In the case of macromolecules at least, a plot of π/c against c has typically the form shown in Figure 1 and the dependence of π on c is

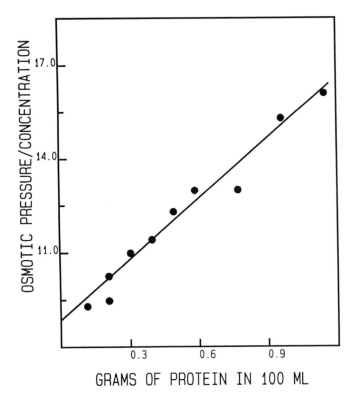

Figure 1. Plot of π/c against c for glycinin, a soya bean protein, in 4 M guanidine hydrochloride showing the importance of the second virial coefficient [B_1 in Eq. (**24**)]. From numerical data of ref. [5], p. 93.

given by a virial expression:

$$\frac{\pi}{c} = RT\left(1 + B_1c + B_2c^2 + \cdots\right) \qquad (23)$$

where the virial coefficients $B_1, B_2, \ldots$ are specific to each system. The integral of Eq. (**21**) takes the form

$$\frac{w}{RT} = \ln[c/c_0] + B_1(c - c_0) + B_2\left(c^2 - c_0^2\right)/2 + \cdots \qquad (24)$$

The activity coefficient γ may be defined by the relation

$$\ln \gamma = B_1(c - c_0) + B_2(c^2 - c_0^2) + \cdots \tag{25}$$

which gives

$$w = RT \ln(\gamma c); \qquad \mu = \mu_0 + RT \ln(\gamma c) \tag{26}$$

The activity coefficient γ tends to unity in the region in which $B_1(c - c_0) \ll 1$. It may differ from 1 at concentrations as low as 0.01 mM for polyelectrolyte macromolecules and 0.01 M for ordinary electrolytes [6]. Uncharged molecules in good solvents may exhibit $\gamma = 1$ at concentrations of 0.1 or even 1 M. Since very few solutions have $\gamma = 1$ at a molar concentration, an activity coefficient of 1 is arbitrarily assigned to the standard concentration c_0 appearing in equations like (21). In the binding of ligands to 0.1 mM protein in solutions which are 0.05 M or more in electrolyte concentration it is customary, and overall correct, to assume $\gamma = 1$. As these conditions are reached in most of the experimental circumstances of interest we shall assume that the general situation is one in which the simpler form of the chemical potential, $\mu = \mu_0 + RT \ln(c)$ may be used.

We note three important properties of chemical potentials:

1. They are extensive quantities that refer, by convention, to 1 mol. A fraction dn of 1 mol has a chemical potential $\mu \, dn$.
2. Of the two terms forming the chemical potential the standard term represents the "true" chemical potential, in the sense that it is this part that determines the chemical behavior of the substance, and can only be known through it. The concentration-dependent term is ideally a universal function of the concentration expressing the osmotic work required to bring the standard to the actual concentration.
3. The standard chemical potentials are not absolute constants. Not only do they depend on temperature and pressure but in principle they depend on *all the other components of the system*. This last circumstance is of limited importance for reactions between very simple molecules to the extent that it receives little comment in texts of chemical thermodynamics. It is customary to define a "standard state" from which, by the simple operations of the potential functions, one may derive the correct chemical potential in media of different composition. Though legitimate in principle this view does not lend itself to practical application in the case of molecules as complex as the proteins. We have to assume that for

them the values of the standard chemical potentials are defined only for media of precise composition and more often than not we are not able to accurately measure, or even estimate, the difference between these values in media of quite close chemical composition (see Chapter II).

Change in Gibbs Free Energy in a Chemical Reaction

The relation and significance of the two parts that make up the chemical potential become clear in describing the free energy change that takes place in a chemical reaction. Let this reaction involve the conversion of n mols of reactants (A and B) into products (C and D); n represents the molar amount transformed, the same for all components. If the initial amounts of the components are $n_A \ldots n_D$, after reaction has taken place A and B have decreased by an amount n and C and D have increased by the same amount:

$$G = \mu(A)(n_A - n) + \mu(B)(n_B - n) + \mu(C)(n_C + n) + \mu(D)(n_D + n)$$

$$(27)$$

From the last equation it follows that

$$\frac{dG}{dn} = \mu_C + \mu_D - \mu_A - \mu_B \qquad (28)$$

or

$$\frac{dG}{dn} = \mu_0(C) + \mu_0(D) - \mu_0(A) - \mu_0(A) - \mu_0(B)$$

$$+ RT\left[\ln(\langle C\rangle\langle D\rangle)/(\langle A\rangle\langle B\rangle)\right] \qquad (29)$$

$\langle A\rangle \ldots \langle D\rangle$ are the *arbitrary* molar concentrations obtaining *while n mol react*. Therefore, the differential coefficient dG/dn on the left-hand side of (28) stands for the finite free energy change dG in the conversion of 1 mol of A and B into 1 mol of C and D *under the condition that the concentrations* $\langle A\rangle \ldots \langle D\rangle$ *remain the same during the conversion.* We shall in the following denote the finite changes in V, S, and G that take place when 1 mol of the components of a system passes from a well-defined initial state to an equally well-defined final state by the symbols ΔV, ΔS, and ΔG, respectively. The term $\mu_0(C) + \mu_0(D) - \mu_0(A) - \mu_0(B)$ in (29) corresponds to the change in Gibbs free energy of the system when molar amounts of A and B are substituted by molar amounts of C and D. It is therefore the standard free energy change ΔG in the reaction $A + B \rightarrow C + D$ and Eq. (29) becomes

$$dG = \Delta G + RT \ln \frac{\langle C \rangle \langle D \rangle}{\langle A \rangle \langle B \rangle} \tag{30}$$

Had we chosen initial amounts $n(A) \ldots n(D)$ of reagents that represent the proportions present when chemical equilibrium is reached no change in Gibbs free energy would have taken place and in consequence $dG = 0$. The arbitrary concentrations $\langle A \rangle \ldots \langle D \rangle$ become then the *thermodynamic equilibrium concentrations* $[A] \ldots [D]$ and

$$\Delta G = -RT \ln \left(\frac{[C][D]}{[A][B]} \right) \tag{31}$$

As ΔG is a constant, at fixed temperature, pressure, and solvent composition, the last equation will be satisfied whenever

$$\frac{[C][D]}{[A][B]} = K, \quad \text{with } K = \exp - \left(\frac{\Delta G}{RT} \right) \tag{32}$$

K is the equilibrium constant of the reaction. Notice that with K and ΔG defined for the reaction $A + B \rightarrow C + D$, the equilibrium constant and standard free energy change for the reaction $C + D \rightarrow A + B$ are respectively $1/K$ and $-\Delta G$. The second term in the right-hand side of (30) represents the difference in the osmotic work required to bring 1 mol of the reactants—on the one hand—and 1 mol of products—on the other—from the unit concentration to the equilibrium concentration. Equation (31) permits calculation of K and ΔG, and therefore of the difference in the *standard* chemical potentials of reactants and products, from the equilibrium concentrations. Notice that the concentrations appearing in Eq. (32) are dimensionless quantities since they are ratios of actual to standard concentration, the latter arbitrarily set to unity. Therefore the thermodynamic equilibrium constant is dimensionless as well, a characteristic made clear by the relation between K and ΔG shown in that equation.

While dG is a differential coefficient valid only when the concentrations of products and reagents are fixed [Eq. (30)], the standard free energy change ΔG is, according to Eq. (32), not subject to that limitation. The constancy of ΔG and K follows only from those of the standard chemical potentials, and these may conceivably vary with the extent of reaction. This possibility is discussed in more detail in Chapter II.

Change in Gibbs Free Energy under Stationary Conditions

Consider Eq. (29) in which we set $dn = 1$. As the concentrations $\langle A \rangle \ldots \langle D \rangle$ are arbitrary concentrations this equation gives the free energy of formation of 1 mol of products from 1 mol of reactants under all conditions in which the concentrations are fixed. Stationary concentrations are operative in a reaction vessel in which A and B are continuously introduced and C and D withdrawn at rates appropriate to maintain the concentrations constant at all times. When the reagents have fixed but otherwise arbitrary concentrations we designate these as $\langle A \rangle$, $\langle B \rangle$, $\langle C \rangle$, and $\langle D \rangle$, a notation designed to distinguish them from the equilibrium concentrations $[A] \ldots [D]$. When the concentrations chosen fulfill the condition (32) they become the thermodynamic equilibrium concentrations and these make $dG = 0$ in Eq. (31). We shall maintain the distinction between arbitrary concentrations $\langle A \rangle \ldots$ and thermodynamic equilibrium concentrations $[A] \ldots$ throughout this work.

Extent of Reaction and Change in Free Energy

If stoichiometric amounts of reagents are employed a quantity e, the extent of reaction, giving the fractional transformation of reactants into products, may be defined [7]. If we set in (27) $n(A) = n(B) = 1$; $n(C) = n(D) = 0$, then $e = dn$ denotes the extent of the reaction that converts A + B into C + D, e being comprised between 0 and 1.

$$\langle A \rangle = \langle B \rangle = 1 - e; \qquad \langle C \rangle = \langle D \rangle = e \tag{33}$$

Equation (29) now reads

$$\frac{dG}{de} = \Delta G + RT \ln\left[\frac{e^2}{(1-e)^2} \right] \tag{34}$$

From this equation,

$$\text{at} \quad e = e_0 \qquad dG/de = 0$$

$$\text{at} \quad e = 0 \qquad dG/de = -\infty$$

$$\text{at} \quad e = 1 \qquad dG/de = +\infty \tag{35}$$

These relations are exhibited in Figure 2. In a spontaneous process G must always decrease, that is $dG/de < 0$ for any e. At equilibrium $e = e_0$

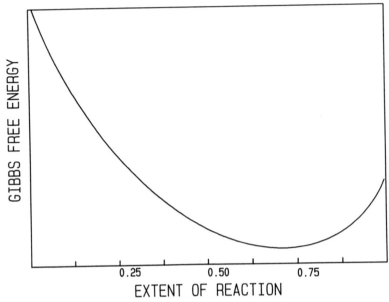

Figure 2. Dependence of G and dG/de on extent of reaction. A standard free energy change $\Delta G = -1.8\ RT$, sufficiently close to zero was chosen, so as to display the minimum in free energy change at a finite extent of reaction. Calculations were made by numerical integration of Eq. (**34**) with e interval of 0.005.

and $dG/de = 0$. If $e < e_0$ de is necessarily positive and the reaction proceeds in the direction A + B → C + D. If $e > e_0$ de is negative and the reaction will proceed in the direction C + D → A + B. Starting at any point on the curve the system moves toward $dG/de = 0$, which occurs at $e = e_0$. We note that the path that leads to equilibrium (Fig. 2) involves a continuous change in the extent of reaction and must therefore be carefully distinguished from the molar free energy change at a fixed extent of reaction [Eq. (**34**)], which corresponds to the differential of this path. In fact, the free energy G corresponding to each value of e in Figure 2 has been determined by numerically integrating Eq. (**34**) between e and a very small initial extent of reaction.

Equilibrium Shift with Temperature and Pressure

Experiment shows that on changing from one temperature or pressure to another there is a shift in the extent of reaction. It has already been seen that

$$\left(\frac{d\,\Delta G}{dp}\right)_T = \Delta V \tag{36}$$

In Eq. (36) ΔV is the "standard volume change" in the reaction and equals the total molar volume of products minus the total molar volume of reactants. For reactions in solution these molar volumes are those of the solvated species. ΔH, the standard enthalpy change in the reaction, is similarly defined as the difference in heat content, at constant pressure, of reactants and products. As already seen the Gibbs–Helmholz equation states:

$$\left\{\frac{d(\Delta G/T)}{d(1/T)}\right\}_P = \Delta H \tag{37}$$

The direction of the changes in the extent of reaction with temperature or pressure is determined by the sign of de/dp and de/dT, respectively. Equation (36) may be written

$$\frac{de}{dp} = \frac{\Delta V}{d(\Delta G/de)} \tag{38}$$

similarly Eq. (37) gives

$$\frac{de}{dT} = -\frac{\Delta H}{T^2 d(\Delta G/T)/de} \tag{39}$$

From Eq. (31), at constant temperature

$$\frac{d\,\Delta G}{de} = -\frac{RT(2-e)}{e(1-e)} \tag{40}$$

and this relation gives an introduction of Eq. (36)

$$\frac{d\,\Delta G/dp}{d\,\Delta G/de} = \frac{de}{dp} = -\frac{\Delta V e(1-e)}{(2-e)RT} \tag{41}$$

Similarly

$$\frac{de}{dT} = \frac{\Delta H e(1-e)}{(2-e)RT^2} \tag{42}$$

As $e(1 - e)$ is always a positive quantity de/dp and ΔV are always of opposite sign while de/dT and ΔH are always of the same sign. If the reaction $A + B \rightarrow C + D$ proceeds with increase in volume ($\Delta V > 0$) then de/dp is negative, that is the reagents that occupy the smaller volume (A and B) are favored. If $\Delta V < 0$, de/dp is positive and the reagents that occupy the smaller volume (C and D) are again favored. By a similar reasoning the change in equilibrium with temperature always favors the direction of a positive enthalpy change (absorption of heat by the system from the surrounding medium).

Practical Characterization of the Chemical Equilibrium

This is best discussed in relation to a simple, obviously easily reversible equilibrium like the proton addition to acids and bases which may be formulated as

$$A + H^+ \rightarrow AH^+ \tag{43}$$

where A designates a charged or uncharged species. If AH^+ is considered as the product, the equilibrium constant is the association constant of the system,

$$K_{ass} = \frac{[AH]}{[A][H]} \tag{44}$$

and if AH^+ is considered as the reactant the equilibrium constant is the dissociation constant of the system

$$K_{dis} = \frac{[A][H^+]}{[AH^+]} \tag{45}$$

The association constant has dimensions of liters/mol and the dissociation constant has dimensions of mol/liter. The physical significance of the dimensions is made clear in the definitions of these equilibrium constants: K_{ass} represents the number of liters of solvent into which 1 mol of complex must be dissolved to obtain dissociation of one-half of the total, while K_{dis} represents the molar concentration of complex at which half-dissociation is present. Since the second number but not the first has direct practical interest we shall adopt the dissociation constant as our standard form of equilibrium constant and will drop the subscript dis in future writing. On the other hand we will find it convenient to use the free energy of *association* of the two ligands to form the complex as our standard form:

$$\Delta G = \mu_0(AH^+) - \mu_0(A) - \mu_0(H^+) \tag{46}$$

With ΔG and K thus defined

$$\Delta G = RT \ln K; \qquad K = \exp(\Delta G/RT) \tag{47}$$

As indicated by Eqs. (47) the choice of the dissociation constant and the association free energy eliminates the negative sign in the relation between these two quantities. The extent of reaction is given by the degree of association or saturation

$$s = \frac{[AH]}{[A] + [AH]} \tag{48}$$

or by the degree of dissociation $\alpha = 1 - s$. Practical recognition of the equilibrium requires determination of the concentrations of all the reagents involved in it. If the experiments determine, as is often the case, α or s and this value differs significantly from both zero and unity one has certainly determined the proportions of A and AH^+ but the concentration of the third partner, H^+, is still to be determined. In the particular case of protons this is easily done: An electrometric procedure is available to measure H^+ independently, even when this is as small as 10^{-14} M. In the typical equilibria of proteins with ligands other than protons one has not at one's disposal such ultrasensitive methods for the determination of the concentration of free ligand. The usual experimental procedure is to mix solutions of protein and ligand of known concentrations and to determine the saturation s with respect to the total binding capacity of the protein. From the degree of saturation and the total concentrations of ligand and protein the amount of free ligand is calculated, so that the uncertainty in the free ligand concentration is wholly dependent on the precision with which s is measured. From this quantity alone we expect to deduce both the dissociation constant of the complex and the stoichiometry, that is the number of moles of ligand bound to the protein at $s = 1$. Presently we examine the case in which all potential binding sites in the protein have equal affinity for the ligand (simple binding, discussed in Chapter IV). Let P_0 and X_0 denote, respectively, the total protein and total ligand in the solution. The dissociation constant is

$$K = \frac{(P_0 - [PX])(X_0 - [PX])}{[PX]} \tag{49}$$

which yields the quadratic equation

$$[PX]^2 - (P_0 + X_0 + K)[PX] + P_0 X_0 = 0 \qquad (50)$$

Equation (50) is symmetric in P_0 and X_0, which can be exchanged without altering the form of the equation. As a consequence s also represents the saturation of ligand with protein. If the saturation of protein with ligand, $s = [PX]/P_0$, $b = 1 + (K/P_0)$, and $y = X_0/P_0$ are introduced in the last equation it becomes

$$s^2 - (b + y)s + y = 0 \qquad (51)$$

If $P_0 \gg K$ then $b \to 1$. With this value of b Eq. (51) reads

$$(y - s)(1 - s) = 0 \qquad (52)$$

the solution of which is evidently $y = s$. Equation (52) is valid when the dissociation constant is sufficiently small in comparison with the protein (or ligand) concentration. In such case the added ligand (or protein) binds stoichiometrically to the protein (or ligand) until saturation is reached. A plot of s against X_0/P_0 according to Eq. (52) shows that as b approaches 1, that is as the reactant held constant increases its concentration beyond the value of the dissociation constant of the complex, s approaches y over an increasingly larger span of s. The region of the plot over which this relation is valid within experimental error is the stoichiometric region in which the added protein, or ligand, appears to be wholly bound to the other partner. Accordingly its free concentration cannot be derived from the value of s. The extent of this stoichiometric region for a fixed value of b is made apparent by writing Eq. (51) in the form

$$y = \frac{s(b - s)}{(1 - s)}; \qquad y = s + s\frac{b - 1}{(1 - s)} \qquad (53)$$

The departure dy from the relation $y = s$ is $s(b - 1)/(1 - s)$. If the method employed permits the determination of s with standard error ds, the critical value of s, $s(dy)$, for which $dy = ds$ is given by

$$s(dy) = \frac{1}{1 + (b - 1)/ds} \qquad (54)$$

Table 2 shows the extent of the stoichiometric region for $d = 0.05$ and increasing values of X_0/K_0. In practice one determines the stoichiometry

Table 2. Titration of Protein at Concentration P_0 with Ligand[a]

P_0/K_0	s(for $ds = 0.05$)
10	0.33
30	0.60
100	0.83
300	0.94
1000	0.98

[a]$s(ds)$ is the saturation at which the departure from linearity in the plot of s against X_0 equals the assumed error in saturation, 0.05.

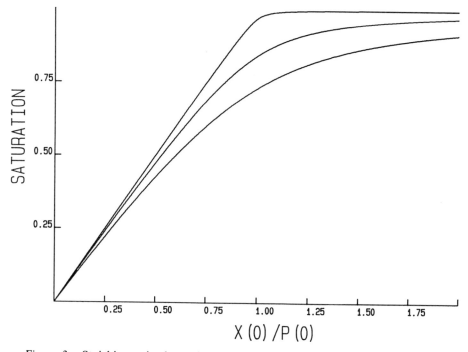

Figure 3. Stoichiometric plot and equivalence point. The three curves, calculated by Eq. (51) are for $b = 0.1$, 0.03, and 0.01, respectively.

of the reaction by choosing for one partner a fixed concentration much larger than the dissociation constant and determines s after addition of increasing amounts of the other partner. Extrapolation of the undisputably linear portion of the plot of s against y to $s = 1$ (Fig. 3) yields the value of y corresponding to $s = 1$, often called the equivalence point.

Precision of the Dissociation Constant

Writing

$$K = [(1 - s)/s](P_0 - sX_0) \tag{55}$$

the fractional error in K is given by

$$\frac{dK}{K} = -ds\left[\frac{1}{s(1 - s)} + \frac{X_0}{P_0 - sX_0}\right] \tag{56}$$

The contribution to the uncertainty by the term in X_0 and P_0 is the more important one when these quantities are greatly different. It becomes prohibitively large as stoichiometric conditions are approached. The incurred error is a minimum if $X_0 = P_0$ and $s = 1/2$ when the second term contributes a factor of 2 and the first, now the more important, amounts to 4, making $dK/K = 6$ ds. If $ds = 0.03$, quite a good precision in s, $dK/K = 0.2$. For s greater than $3/4$ or smaller than $1/4$ the imprecision increases rapidly. Following these considerations it is not surprising to find that few values of the dissociation constants of protein–ligand complexes quoted in the biochemical literature have a precision better than 50%. Measurements of stoichiometry are best accomplished under conditions in which the precision of measurements of K is far from optimum and reciprocally K must be measured well away from conditions of stoichiometric addition of protein and ligand. As a consequence the plots in which, by means of an appropriate extrapolation, both stoichiometry and dissociation constant are simultaneously obtained must do so at the cost of precision in both these quantities.

References

1. No more appropriate, or insightful, introduction to the general subject of thermodynamics is available than Pippard, A.B. (1964) *The Elements of Classical Thermodynamics*. Cambridge University Press, Cambridge. Equally valuable is Sommerfeld, A. (1964) *Thermodynamics and Statistical Mechanics*. Academic Press, New York. Presentations of thermodynamics that can serve

as introduction to the chemical and biochemical applications are many. The classical one is that of Lewis, G.N., and Randall, M. (1923) *Thermodynamics, and the Free Energy of Chemical Substances.* McGraw-Hill, New York. Other useful expositions are Zemansky, M.W. (1957) *Heat and Thermodynamics.* McGraw-Hill, New York, and Glasstone, S. (1947) *Thermodynamics for Chemists.* van Nostrand, New York.

2. The chemical potential is often introduced formally as the partial differential of the free energy with respect to the molar fraction of the ith component $\mu_i = dG/dm_i$ (e.g., Zemansky 1957, p 420). I have chosen to define it as determined by the empirical chemical equilibrium and its dependence on the concentration of the reagents.

3. Emden's famous statement in *Nature* (*London*) **141**, 908 (1938) "we require heating in winter because we need entropy rather than energy" sounds quite natural to the biologist: If the temperature is sufficiently low we shiver and lose chemical energy to gain entropy.

4. The law of mass action (Guldberg and Waage, 1867) and van't Hoff (1877) is, in the words of Lewis and Randall, "one of the milestones in the progress of chemistry as an exact science."

5. The osmotic pressure of proteins has had great importance as the original method for the determination of their molecular weights. For a detailed exposition see Tombs, M.P., and Peacocke, A.R. (1974) *The Osmotic Pressure of Biological Macromolecules.* Clarendon, Oxford, Parts 1 and 4.

6. Activity coefficients as corrections to the concentration in the chemical potential equation were introduced by Lewis and Randall [1], Chapter XXII. For the relation to osmotic pressure deviations see Glasstone [1], Chapter XIV.

7. Discussed more fully in Zemansky [1], Art. 17.12.

II

The Chemical Potentials of Proteins

The Concept of Equilibrium in Theory and Practice

We have defined a state of equilibrium as one in which the change in free energy with extent of reaction has reached its minimum, but such a definition has little practical value as an experimental criterion to decide when equilibrium has been attained and when we are away from it. In practice a state of the system is defined by the value of one or more experimental quantities. The persistence of these quantities for an arbitrary length of time is certainly a necessary, but not a sufficient indication that equilibrium has been attained. In a chemical reaction a sufficient condition appears to be that the same final composition be achieved when starting with either excess reactants or excess products, since it is only at equilibrium that $dG/de = 0$. In practice application of this criterion is not always a simple matter. In many reactions the equilibrium point corresponds to such preponderance of products over reactants that reversibility is difficult or seemingly impossible to demonstrate by starting with "pure" products. In an equilibrium of protein with ligand it is easy to follow the dissociation of the complex by dilution, but far less easy to provide for complex formation by concentration. The state of "indefinitely persistent equilibrium" is an idealization never realized in practice: A reaction may reach actual equivalence of opposing chemical processes but these will be *always* followed—or accompanied—by others, even if the results of the latter reactions are only noticed a very long time after apparent completion of the most prominent one. This circumstance arises because the system under study is, in reality, never isolated but forms part of a larger system that evolves toward its own state of equilibrium over a longer time. In deciding that thermodynamic equilibrium has been practically reached we disregard any subsequent or concomitant processes, by assuming that their influence on the reaction under study is so small that the thermodynamic parameters calculated from the observations will differ by an arbitrary small amount from those that pertain to an indefinitely stable, ideal equilibrium. As a consequence the kinetic characteristics of the approach to equilibrium provide the most useful, and in many cases the only practical criterion for the demonstration of the existence of an equilibrium. The difference in the rates of formation of products and disappearance of reactants approaches zero as the instantaneous composition of the system approaches the equilibrium point. As a general rule if the rate of measurable reaction halves in time t^* one can expect to be sufficiently close to the equilibrium point in a time of $4t^*$ to $6t^*$. In uncatalyzed reactions t^* may be prohibitively long but the chemical reactions of physiological significance are, of necessity, accomplished in short times. If the states thus reached appear to be subsequently stable over times that are biologically relevant it is useless to speculate on whether they correspond to "true" thermodynamic equilibria or whether this is still to be achieved by surmounting an additional potential barrier. With the reservations explained above we can treat the state thus reached as one of "true" thermodynamic equilibrium.

The Chemical Potentials of Proteins

Classical chemical thermodynamics was developed from observations on simple chemicals endowed with virtually indefinite stability. Their chemical reactions, almost without exception, transform them into well-defined products of similar stability and the theory of chemical potentials that we have outlined in Chapter I can be applied to them without reservations. We may note that at the time of the proposal by Gibbs of the concept of chemical potential (1876) neither flexible polymer molecules nor transient molecular complexes were known and that even electrolytic dissociation was only dimly understood. The macromolecules, and in particular the proteins, have a compositional and conformational complexity that makes the definition of chemical potentials much more involved, and the application to proteins of the rules that prove satisfactory for simpler systems is often anything but straightforward. We recall that we are interested in the changes of the thermodynamic potentials and of the entropy and that it is the difference between values of these quantities that is of practical interest, while their absolute magnitudes are immaterial. In the particular case of macromolecules the only certain means for the determination of differences in chemical potentials is through the observed equilibrium constants of their chemical reactions. A protein possesses a variety of reacting modes and what we come to derive from the chemical reactions of the protein are the chemical potentials associated to each particular reacting mode. By extension, we can speak of the "chemical potential of the protein" as a quantity resulting from the superposition of the chemical potentials of its various parts, although such definition has no appreciable practical value. When we speak of the chemical potential of a protein it is always with reference to some particular reaction, e.g., dissociation of a proton, binding of a ligand, redox potential of a group like S–S, heme, flavin, etc. The reactivity of groups in the protein is influenced by the surrounding protein elements so that the chemical potential depends, to a variable extent, on the three-dimensional conformation of the protein. The influence of the solvent composition is not a trivial one, as it results not only from direct effects of solvent components on the reacting groups but equally often from solvent-induced changes in the protein conformation. Small molecules can also exist in several conformations in equilibrium (e.g., boat–chair, keto–enol) and it is well understood that their chemical potentials result from the sum of the molar chemical potentials of the conformations in which they exist each weighted by the mol fraction that it represents, in accord with Eq. (I.11). Spectroscopic experiments (NMR, fluorescence, Mössbauer spectroscopy) and isotopic hydrogen exchange have indicated that proteins undergo structural fluctuations extending into the nanosecond, and even shorter time ranges. The protein in solution must then be considered as a population of conformers $P_1, P_2 \ldots$ in proportions $f_1, f_2 \ldots$ with respective free energies $G_1, G_2 \ldots$. The average chemical potential of the protein, $\langle G \rangle$—always in relation to some particular reaction mode—is

$$\langle G \rangle = G_1 f_1 + G_2 f_2 + G_3 f_3 + \cdots \tag{1}$$

The fluctuations that are responsible for the presence of the conformers also ensure that the protein will assume a new conformer distribution when the solvent composition, temperature, or pressure are changed. Present experimental data indicate that the distributions to which Eq. (1) refers are virtually "compact," that is, they do not correspond to two or more conformations separated by a free energy gap or gaps that are large in comparison with the thermal energy. Experiments that indisputably prove the existence of protein isomers in solution comparable to the case of boat–chair or *cis–trans* isomers of small molecules are not known to the author. Starting with Monod *et al.* [1] it has been asserted that in the absence of any external heme–ligand hemoglobin exists in two very different conformations, T and R, present in proportions of 1 to $\sim 10^{-4}$, but over these many years the proof of the coexistence of these conformations has not been forthcoming. Similar equilibria between protein isomers have been postulated but never directly demonstrated in schemes of ion transport by the Ca^{2+} and the Na^+–K^+-ATPases [2]. We can definitely assert that for the purpose of the energetics we need not explicitly consider the existence of such separate isomers. If they indeed exist, and this has to be proven in each particular case, we can ignore them: Eq. (1), that states the single-valued character of the average chemical potential of each chemical species, is sufficient to deal with the thermodynamic description of the system. Accordingly we shall adopt—at least as a starting viewpoint—that a protein in a *medium of defined composition* has a unique chemical potential, or rather a set of unique chemical potentials associated with its possible modes of reaction and that *these can be modified only by the covalent or noncovalent binding of ligands.*

Persistent Metastable States

We must note, however, the possible creation of metastable states through a series of cyclic operations: On changing the solvent composition from A to B, or after a change in temperature or pressure, the chemical potential of some protein groups may change and it may be that on reverting to medium A the conformation, or the state of aggregation, and therefore the chemical potentials do not return *immediately* to their original A-characteristic value. We have thus created a metastable state and reaction in medium (conditions) A will take place with a decrease in free energy that more nearly corresponds to the reaction in medium (conditions) B. This may be seen as conceding, in opposition to the previous statements, that a protein dissolved in a well-defined medium can have a multiple-valued chemical potential depending on previous history. This apparent contradiction disappears if we remember that the metastable chemical potential has been obtained as a result of transfer from one medium (A) to another (B), or by a reversible change in temperature or pressure. Because free energy is conserved, the difference in chemical potentials in medium A before its transfer to B and after return to it derives from the free energy input required for transfer of the protein from one medium to the other. Metastable chemical potentials may then be postulated in the explanation of certain chemical reactions of the protein if an external free energy source is apparent, and is judged to be capable of providing the additional free energy. This criterion may be extended to explain the variety of behaviors that a protein may exhibit depending on the methods used in its isolation: There is no certainty that, as has been often assumed, such differences result always from artifacts derived from the experimental procedures employed rather than from intrinsic properties of the material.

Can Protein Equilibria Be Considered as Totally Time Independent?

In attempting to assign weights to the particular conformations present in a medium of known composition, at a definite pressure and temperature [Eq. (1)] we find that in the case of proteins, complications arise that are not encountered with simpler molecules. To clearly define these consider first in detail the equilibrium of a molecular complex AB with the component molecules A and B. If the spatial conformation of the free partners differs from the conformation that they adopt in the complex it follows that changes in A and B must take place following the formation of AB and revert on dissociation of the complex. These changes cannot be "infinitely fast" and therefore in attempting to apply Eq. (1) we find that the proportion of free partners (A and B) having conformations that approach in some respect the one that they have in AB will steadily increase with the degree of association. The value of $\langle G_A \rangle$ or $\langle G_B \rangle$ will change with extent of reaction, apparently contradicting the statement that for any given substance the chemical potential has a fixed value once the variables of state and the chemical composition of the medium are specified. This contradiction is only apparent as the composition of the medium is indeed changing with the extent of reaction, A and B finding themselves in the presence of increasing concentrations of AB, but in classical chemical thermodynamics an implicit exception to this rule has been customarily assumed by asserting that chemical potentials are independent of the extent of reaction. If we now consider that the changes in conformation on association and dissociation, though not instantaneous, are completed in times that are negligible in comparison with the time of an association–dissociation (AD) cycle we conclude that $\langle G_A \rangle$ and $\langle G_B \rangle$ though not absolutely constant will vary so slightly with the degree of dissociation that no great error will be incurred if they are considered to have unique values. This is indeed the case with complexes of small molecules: The free energies of formation of weak molecular complexes are only of the order of 1–3 kcal and equilibria involving detectable amounts of complex and free ligands occur in the range of 10^{-1}–10^{-3} M. The time for an AD cycle is in these cases determined by the rate of association and is therefore in the range of microseconds. In contrast the expected small changes in the molecular conformations may not take more than a fraction of a nanosecond. In consequence we can consider that the contribution to the conformations of the complex and free partners from the previous state of the molecule, respectively dissociated or associated, will be negligible at all degrees of dissociation. This will not be the case with proteins, Observations of the slow regain of catalytic activity and actual physical parameters of oligomeric enzymes on reversal of dissociation, as well as the dependence of the dissociation constant of certain protein dimers on the degree of dissociation, show that the changes in conformation that follow the association and dissociation of protein oligomers can take a time of the same order, or even longer than that of the corresponding AD cycle. In these cases we expect appreciable variation in the chemical potentials of reactants and products with the extent of reaction. We shall examine in detail the observations that lead to this conclusion in dealing with the equilibria between protein subunits and their specific aggregates (Chapters XIV and XV). At present we shall simply point out two important consequences:

1. As changes in conformation of the aggregate take place *after* the association of the free partners, and equally changes in conformation of the free partners *follow* the dissociation of the aggregate it is evident that there is a prescribed succession of changes and therefore a time arrow during an AD cycle, not only at all stages during the approach to equilibrium but also when equilibrium—as operationally defined above—is reached.

2. We can now define the chemical potentials of a substance as depending on the concentrations of all the components of the system, including those that are formed in their chemical reactions. Moreover this need not arise as a trivial consequence of a further chemical equilibrium between the reactants and the formed product but as the results of time-dependent effects. The proposition that such effects are to be discounted when a "true" equilibrium is established was reached because the analysis was limited to very simple systems in which the changes in conformation following association and dissociation could be disregarded.

The Conformational Space of Proteins

In the classical phase space a single point in N-dimensional space is representative of the average coordinates and momenta of the system. We can adapt this notion to permit a representation of the states that make up the average conformation of the protein. Since momenta are not pertinent to our purpose, our N-dimensional space is one of coordinates alone, with origin at the *average* value of the coordinates of whatever protein features are supposed to be relevant, at a fixed pressure and temperature. A conformational fluctuation will be represented by a point at some distance from the origin, and this distance will bear a direct relation to the free energy difference with the average. If we consider an elementary fluctuation, that is a fluctuation of one coordinate value corresponding to a change in free energy of one thermal unit we will find that a fraction $\exp(-1) = 0.36$ of the molecules occupy a point in the conformation space that no longer coincides with the average. If there are N such relevant independent possibilities of fluctuation we expect the fraction of points that no longer coincides with the average to be $p = Ne^{-1}/(1 + Ne^{-1})$. Thus for a value of $N = 10$, $p = 0.78$ and for $N = 20$, $p = 0.88$. The lack

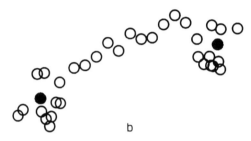

Figure 1. Conformational space of proteins. The centers of the circles correspond to points in N-dimensional space. The dark circles are the minimum energy conformations. The swarm of random points around (open circles) represent states of higher energy that are long-lived with respect to the times for equipartition of energy (picoseconds). (a) Single state. (b) Two states in equilibrium connected by a trajectory of intermediate states of unknown population.

of coincidence with the average arises from differences in a single coordinate, all others being equal to their respective average values. A graph of the representative points that differ from the average would determine an N-dimensional volume surrounding it (Fig. 1a). In an experiment like X-ray diffraction, which takes a very long time in comparison with the rate of fluctuations, all coordinates coincide with their average values over most of the experimental time, and they contribute primarily to the diffraction pattern from which the crystal structure is derived, while their fluctuating values result in some diffuse noise. The speed of purely energetic fluctuations, like those that result from increased vibrational energy, is very large. From the experiments on the decay of vibrational states employing laser spectroscopy [3] we know that they do not last beyond a few picoseconds. They must be clearly separated from energy-dependent structural fluctuations, which though much rarer, must decay at much slower rates. Formation and decay of structural fluctuations must be expected to proceed against the frictional resistance offered by neighbor structures. A variety of experiments [4], a well as the simulation of the protein fluctuations by the method of molecular dynamics [5] has provided evidence that displacements of structures within the protein take place facilitated by the very short-lived energetic fluctuations. In some manner, which is not known to us presently in any detail, these fluctuations must eventually permit the larger changes of structure that are connected with the dissociation of oligomeric proteins (Chapter XIV) and other phenomena in which changes in protein conformation have been postulated. A conformational change between two stable states consists of a trajectory such as indicated in Figure 1b: the two average states correspond to single points surrounded by others that indicate the possible structural conformations that arise from relatively stable fluctuations of the respective averages. The conformation entropy must be related to the density of such points. If each average state comprises N_1 and N_2 states of identical probability we expect the entropy associated with the transition from 1 to 2 to be $\Delta S = k \ln(N_2/N_1)$. However, just like in the case of X-ray diffraction, the characteristic time of the experimental technique employed to study the equilibrium determines the number of fluctuation states that are "seen" by it, and therefore the entropy that would be computed. If the method used has a characteristic time τ_c and the N_1 and N_2 conformations within each group exchange within themselves in time τ_x we expect the computed entropy change to be $\Delta S = [\tau_c/(\tau_c + \tau_x)]k \ln(N_2/N_1)$. Even if we know τ_c we may not find much use for the last equation because of our almost complete ignorance of τ_x, but it helps us to understand why different methods may yield different values for the thermodynamic parameters. Moreover, we can surmise the existence of different conformations in equilibrium through the changes of ΔS with temperature, as we expect τ_x to decrease with increase in temperature.

Entropy and Enthalpy Changes Linked to Protein Conformation

The original formulation of thermodynamics attempted to interpret the limitations of the conversion of thermal energy that flows between bodies at different temperatures into mechanical work. In time it was realized that the concepts thus derived were of general value, in particular in the conversion of chemical energy into other forms. The concepts of chemical potential and free energy of reaction by Gibbs aimed at adapting them to the new interests. It is not, however, a simple matter to determine in any given case the microscopic molecular phenomena that are responsible for the separate changes in enthalpy and entropy. If we consider a system formed by molecules of two types, like solvent and solute, we expect that the formation of bonds between them in any isothermal process will decrease the number of possible conformations in the system, so that a decrease in enthalpy will necessarily imply a decrease in entropy and similarly the disappearance of some of these bonds will lead to increases in enthalpy and entropy. Thus, enthalpy and entropy increases and decreases may be expected to be linked. It is evident that most of the possible decrease in entropy that follows bonding between two structures will require a limited enthalpy decrease. Bonds of greater enthalpy will result in the same entropy decrease and a free energy of association that is larger in absolute value. It is a characteristic of biological systems, brought about by the necessity of their regulation, that the free energy of stabilization of almost any noncovalent equilibrium state be small, rarely exceeding 2 or 3 kcal. In the case of complexes of small molecules the changes in bond enthalpy estimated from the known structures, or determined by experimentation, commonly exceeds the experimental free energy change by a considerable amount. It follows that the small free energy must result from an entropy decrease only smaller in magnitude than the enthalpy. Larger changes in free energy, e.g., those that occur on the formation of protein aggregates, or in the folding of an extended peptide require changes in a much larger number of bonds. Changes in enthalpy and entropy presuppose that heat is absorbed from or released to the medium, but only the former (heat absorption) increases the number W of possible molecular conformations that determine the increase in entropy according to the well known Boltzmann relation, $\Delta S = k \log \Delta w$.

Proteins are compact polymers cross-linked at a large number of atoms by forces that are weak in comparison with the covalent bonds. A completely unfolded peptide chain in which all internal interactions are broken would have maximum enthalpy and entropy. Folding of the chain owing to the interactions characteristic of the native globular protein results in large decreases both of enthalpy and entropy. A rise in temperature produces the breaking, or weakening, of some of the internal chain bonds and the result is an increase in the number of spatial conformations present and therefore an increase in entropy. Over the limited range of temperatures in which the protein is functionally competent there is an approximately linear correlation [6] of the changes in entropy and enthalpy that result from binding of a ligand, and therefore a partial compensation of their contributions to the free energy. Also a temperature of perfect compensation can exist at which $\Delta G = 0$. We may expect that this temperature is close to that of best function, simply from the fact that a poised system that must respond to changes in additional components by a moderate displacement of the equilibrium can never be very far from the condition $\Delta G = 0$. While it is in principle possible to enumerate the contributions to the entropy by the changes in the degrees of freedom in small molecules, and to attempt an estimate of these changes on complex formation, this task appears close to impossible in the case of proteins. At present we lack even crude estimates of the residual entropy of proteins and its temperature changes under standard conditions and we have no methods by which we can estimate the importance of protein entropy changes from the overall entropy change on the binding of ligands.

Albeit crudely we can summarize our scant knowledge of this area by the statement that changes in the internal free energy (i.e., stability) of the protein are buffered by the enthalpy–entropy compensation.

Native Conformation, Denaturation, and Chemical Potential of Proteins

Exposure of globular proteins to acid conditions, to high concentrations of urea or guanidine, to pressures of several kilobars or to temperatures of 50 to 90°C often results in the appearance of a modified fraction of the protein that remains insoluble at ambient temperature and pressure in solvent media in which the untreated protein was freely soluble. This phenomenon of "denaturation" was once believed to provide proof that proteins are physically metastable entities whose spatial conformation is irreversibly destroyed by the agencies mentioned. However, a detailed examination of many particular cases has shown that if no changes in covalent bonds occur as a result of the denaturing treatment, conditions can be found in which the native properties are regenerated. Hydrodynamic and spectroscopic methods provide evidence that under denaturing conditions the proteins lose their native compact globular shape with a great increase in the number of non-hydrogen-bonding (hydrophobic) amino acid residues exposed to the solvent. Aggregation of the unfolded protein molecules through these exposed solvophobic groups [7] is the main obstacle that has to be overcome to regenerate the native conformation on reverting to the original conditions. Isolation of the molecules from each other by dilution, by increase in electrostatic repulsion, or by attachment of certain ligands [8] can be used to bring about the separation of the molecules from each other with recovery of the characteristic native properties. It is concluded from these observations that the isolated molecule of protein will, by the operation of its own dynamics, revert back to the native form. Following its synthesis in the ribosomes the polypeptide is in similar isolation from other molecules and assumes rapidly its characteristic native conformation. The question of whether this is the conformation with the lowest free energy or the one that most readily forms and persists cannot be answered and in fact presents little interest. The assignment of a unique chemical potential requires only the persistence of the average conformation over the time necessary to carry out the particular experiments, or the biological functions, that are to be interpreted on the assumption of a single-valued chemical potential. One should recall in this respect that a protein in water solution is already a chemically metastable system as the addition of an appropriate protease brings about its hydrolysis to the component amino acids.

References

1. Monod, J., Wyman, J., and Changeux, J.-P. (1965) *J. Mol. Biol.* **12**, 88–118.

2. Post, R.L., Sen, A.K., and Rosenthal, A.S. (1965) *J. Biol. Chem.* **240**, 1437–1445. Fahn, S., Koval, G.J., and Wayne Albers, R. (1966) *J. Biol. Chem.* **241**, 1882–1889. Moczydlowski, E.G., and Fortes, P.A.G. (1981) *J. Biol. Chem.* **256**, 2346–2356, **256**, 2357–2366.

3. Lauberau, A., and Kaiser, W.A. (1975) *Annu. Rev. Phys. Chem.* **26**, 83–89.

4. Gurd, F., and Rothberg, T.M. (1979) *Adv. Protein. Chem.* **33**, 73–165. Debrunner, P.G., and Frauenfelder, H. (1982) *Annu. Rev. Phys. Chem.* **33**, 283–299.

5. McCammon, J.A., Gelin, R.R., and Karplus, M. (1977) *Nature (London)* **267**, 585–590.

6. A formal exposition is given by Lumry, R., and Gregory, R.B. (1986) In *The Fluctuating Enzyme*, G.R. Welch, ed. Wiley-Interscience, New York, pp. 58–77. Connelly, P., Ghosaini, L., Hu, C.-Q., Kitamura, S., and Sturtevant, J.M. (1991) *Biochemistry* **30**, 1887–1891. Sturtevant J.M. (1987) *Annu. Rev. Phys. Chem.* **38**, 463–488.

7. Barbu, E., and Joly, M. (1952) *Faraday Soc. Discuss.* **13**, 77–93 First demonstration of the reversible aggregation of denatured protein.

8. Gibbons, I., and Perham, R.N. (1970) *Biochem. J.* **116**, 843–849. Klotz, I.M. (1967) *Methods Enzymol.* **11**, 576–580.

III

Simultaneous Equilibria of a Protein with Several Ligands

The chemist who investigates the molecular complexes of small molecules usually has to contend with equilibria in which ordinarily no more than two components, besides the solvent, are involved. Not so the protein chemist: proteins are polyelectrolytes and they cannot be found in aqueous solutions free from counterions. Their properties are critically dependent on the ionic strength and specific ionic composition of the solvent, particularly the concentration of hydrogen ions. The simplest protein system encountered in practice is composed of the protein molecule interacting with the solvent and at least two kinds of ions, besides any added specific ligand. We can well imagine the oversimplification involved in considering such system as virtually comprising a single component in solution, "the protein," the chemical potential of which determines its energetics without reference to any internal complexity. We must accept from the start that when a ligand is added to the solution containing the protein we never deal with one single ligand–protein equilibrium but instead that simultaneous interacting equilibria are the rule in these cases. We need then no apologies for introducing the simultaneous binding of two different ligands as the simplest, as well as the standard case [1]. Those phenomena sufficiently simple to be explained on the "one protein–one ligand" assumption will then appear as limiting cases that will very likely become rarer as experimental methods improve and smaller degrees of interaction between ligands become detectable.

Binding of Two Ligands, X and Y: Free Energy Coupling

Consider the chemical reactions listed I to IV below:

$$
\begin{array}{lc}
\text{I} & \text{P} + \text{X} \rightarrow \text{PX} \\
\text{II} & \text{P} + \text{Y} \rightarrow \text{YP} \\
\text{III} & \text{PX} + \text{Y} \rightarrow \text{YPX} \\
\text{IV} & \text{YP} + \text{X} \rightarrow \text{YPX}
\end{array}
\tag{1}
$$

The standard free energy changes in Reactions I and II will be designated here as $\Delta G(X)$ and $\Delta G(Y)$, respectively. Reactions III and IV correspond, respectively, to the binding of Y when X is already bound and to the binding of X when Y is already bound. Their standard free energy changes will be written $\Delta G(Y/X)$ and $\Delta G(X/Y)$, respectively. They will be called "conditional standard free energies of binding" or simply "conditional free energies" [1]. They stand in contrast to $\Delta G(X)$ and $\Delta G(Y)$, the unconditional free energies, which presuppose the absence of the second ligand. The same final product, YPX, is formed as a result of the successive addition of two ligands: Reaction I followed by III or Reaction II followed by IV. We can apply the principle of free energy conservation in the form:

$$\Delta G(X) + \Delta G(Y/X) = \Delta G(Y) + \Delta G(X/Y) = \Delta G(XY) \qquad (2)$$

$\Delta G(XY)$ designates the standard free energy change in the formation of YPX from the reactants Y, P, and X in their standard states. There is evidently no requirement that $\Delta G(X) = \Delta G(X/Y)$ or that $\Delta G(Y) = \Delta G(Y/X)$ but the conservation of free energy (see Chapter I) imposes the condition

$$\Delta G(X/Y) - \Delta G(X) = \Delta G(Y/X) - \Delta G(Y) = \delta G_{xy} \qquad (3)$$

Thus, the difference between the unconditional free energy of binding for X and its conditional free energy when Y is bound must be the same as the difference between the unconditional free energy of binding for Y and its conditional free energy when X is bound. In other word, the free energy conservation condition leads directly to reciprocity between X and Y, in the sense that the binding of X produces on the binding of Y the same effects that the binding of Y produces on the binding of X. The unique difference between conditional and unconditional free energies for the binding of X or Y will be called the standard free energy coupling between X and Y in the complex YPX. It will be symbolized by δG_{xy}. From Eqs. (1) and (3) it is readily found that

$$\delta G_{xy} = \Delta G(XY) - \Delta G(X) - \Delta G(Y) \qquad (4)$$

The free energy coupling is therefore also the difference between the free energy of formation of the ternary complex YPX and the sum of the free energies of formation of the binary complexes PX and YP. The relations between the various standard free energies of formation of the complexes and the free energy coupling between the ligands is graphically shown in

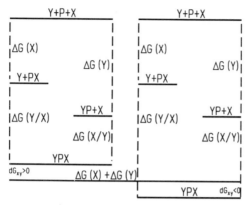

Figure 1. Free energy level scheme showing negative $(\delta G < 0)$ and positive $(\delta G > 0)$ free energy couplings.

Figure 1 as a "Gibbs energy level scheme," a device that we shall often utilize. In general δG_{xy} may be equal to zero or be greater or smaller than zero. When $\delta G_{xy} = 0$ there is no interaction between the bound ligands and the free energy of formation of the ternary complex is the sum of the free energies of formation of the binary complexes. If $\delta G_{xy} <> 0$ its sign is determined by the definition in Eq. (3), which indicates that $\Delta G(X)$ is to be subtracted from $\Delta G(X/Y)$ or $\Delta G(X) + \Delta G(Y)$ subtracted from $\Delta G(XY)$, and not the reverse. Without much loss in generality it may be assumed that $\Delta G(X)$, $\Delta G(Y)$, $\Delta G(X/Y)$, $\Delta G(Y/X)$, and $\Delta G(XY)$ are all negative quantities. It then follows that if $\delta G_{xy} > 0$ there is antagonism between the ligands, which then oppose each other's binding, but if $\delta G_{xy} < 0$ there is cooperation between the ligands, so that they facilitate each other's binding.

Chemical Interpretation of the Free Energy Coupling

Consider the change in standard chemical potentials involved in the unconditional and conditional binding reactions, I and III, respectively:

$$\Delta G(X) = \mu^0(PX) - \mu^0(P) - \mu^0(X) \tag{5}$$

$$\Delta G(X/Y) = \mu^0(YPX) - \mu^0(YP) - \mu^0(X) \tag{6}$$

From Eq. (3) it follows that

$$\delta G_{xy} = \mu^0(YPX) + \mu^0(P) - \mu^0(PX) - \mu^0(YP) \tag{7}$$

From the last relation it follows that δG_{xy} is the standard free energy change in the disproportionation reaction

$$PX + YP \rightarrow P + YPX \tag{8}$$

with equilibrium constant

$$K_{xy} = \frac{[YPX][P]}{[YP][PX]} \tag{9}$$

If $\delta G_{xy} < 0$ then $K_{xy} > 1$ and YPX and P predominate over PX and YP. The binding of the two ligands being cooperative Y and X are more likely to be bound to the same molecule than occupying separate molecules of protein. The opposite occurs if $\Delta G_{xy} > 0$. In this case PX and YP predominate: as a result of the antagonism between the bound ligands these are more likely to be found in different protein molecules. To compare the ligand distributions that obtain as a function of δG_{xy} we consider in detail the situation that arises when the free ligand concentrations, [X] and [Y], are adjusted so that each ligand occupies one-half of the sites that it can potentially fill. Then the overall saturations of the protein for X and Y, which we call respectively s_x and s_y will equal $1/2$. If the total protein concentration is P_0

$$s_x = \frac{[PX] + [YPX]}{P_0} \tag{10}$$

$$s_y = \frac{[YP] + [YPX]}{P_0} \tag{11}$$

$$s_{xy} = \frac{[YPX]}{P_0} \tag{12}$$

$$s_x = s_y = 1/2$$

With these substitutions Eq. (9) becomes

$$K_{xy} = \frac{s_{xy}^2}{\left(1/2 - s_{xy}\right)^2}; \quad s_{xy} = \frac{1}{2} \frac{\sqrt{K_{xy}}}{1 + \sqrt{K_{xy}}} \tag{13}$$

For $K_{xy} = 1$, $s_{xy} = 1/4$ as expected for the random distribution, which is operative if each ligand binds independently of the other. If $K_{xy} <> 1$

the ligands are no longer distributed at random among the protein molecules. From the last equation

$$\frac{-\delta G_{xy}}{4.604RT} = \log\left(\frac{2s_{xy}}{1 - s_{xy}}\right) \tag{14}$$

A plot of δG_{xy} against s_{xy} according to Eq. (14) reproduces an ordinary titration curve [2] (Fig. 2). Zero free energy coupling corresponds to $s_{xy} = 1/4$. For values of s_{xy} symmetrically disposed about 1/4, δG_{xy} has equal absolute values of opposite sign becoming infinite when s_{xy} equals 0

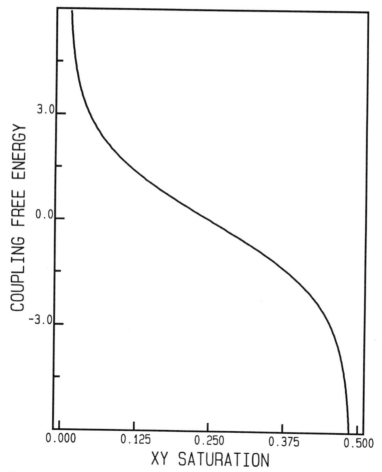

Figure 2. Plot of the double saturation [S_{xy} of Eq. (12)] against dG_{XY}.

or $1/2$. If $T = 300$ K and $\delta G_{xy} = +2.76$ and -2.76 kcal, respectively, the corresponding values of s_{xy} are 0.045 and 0.455, respectively. For $\delta G_{xy} = +2.76$ kcal X and Y are bound to different protein molecules in 90% of the total cases, while for $\delta G_{xy} = -2.76$ kcal X and Y find themselves bound to the same protein in 90% of the cases. Thus for a free energy coupling of this small absolute value we obtain a very good, though not yet perfect, correlation in the binding of both ligands. Figure 2 shows that for an absolute value of δG_{xy} of ~ 2.5 kcal. the plot of s_{xy} against δG_{xy} presents almost an inflexion so that beyond this value of s_{xy} the free energy coupling required to enhance the ligand correlation increases steeply. We encounter here a situation that presents itself regularly in considering the biological properties of proteins: Certain functions will be preferentially associated with the singly or the doubly liganded species. In these cases the perfection of the correlation will be directly related to the efficiency with which the associated biological function can be carried out. A modest free energy coupling of magnitude 1–2 kcal. would be all that is required to achieve a satisfactory correlation while a perfect correlation could be achieved only by a free energy coupling of infinite magnitude. Very often the experimentalist, in an attempt to visualize the effects, postulates a "mechanical" molecular model that virtually ensures a perfect correlation between the binding of two ligands. Evidently such perfect mechanical contrivances are ruled out by the infinite free energy couplings that they theoretically require, as well as by the modest values of δG_{xy} encountered in practice. It is remarkable that this far-reaching consequence as regards the unsuitability of mechanical models of molecular behaviour may be deduced, without any kind of doubt, from the very simple properties of the chemical potential.

Free Energy Couplings and the Linkage Concept

Henderson [3] (1920) appears to have been the first author who realized the thermodynamic implications of the binding of two ligands, which in his case were oxygen and carbon dioxide while the protein was hemoglobin (Fig. 3). Wyman [2] generalized these facts by introducing the concept of thermodynamic linkage. He gave it a form more general than the one I employed above, by stating the equality of the differential coefficients of the changes in chemical potential of each ligand, $\mu(X)$ or $\mu(Y)$, with respect to saturation of the protein with the other ligand, S_y or S_x, respectively:

$$d\mu(Y)/d\mu(PX) = d\mu(X)/d\mu(YP)$$

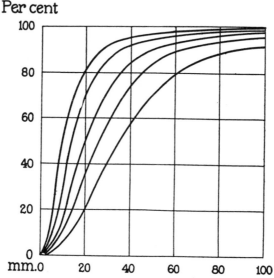

Figure 3. The earliest example of free energy coupling between ligands (O_2 and CO_2) bound to a protein (hemoglobin). Plot of percentage saturation against O_2 pressure in mm of Hg. From left to right the curves are for 0, 3, 20, 40, 90 mm of Hg of CO_2. Observation and blood of Joseph Barcroft. From ref. [2].

or

$$d \ln[Y]/dS_X = d \ln[X]/dS_Y \qquad (15)$$

In spite of its general character this elegant formulation is not of immediate practical value because the differential coefficients that it introduces are difficult to determine experimentally and are subjected to much larger experimental uncertainties than the integral quantities $\Delta G(X)$ and $\Delta G(X/Y)$. Additionally the formulation in terms of free energy couplings between bound ligands has a direct chemical representation in terms of the individual protein complexes, which is not immediate in Wyman's original treatment.

Experimental Demonstration of Reciprocity of Free Energy Coupling between Ligands

Although there are plenty of observations in the literature that reveal the qualitative aspects of the reciprocal effects of bound ligands very few observations have been made with the express purpose of verifying it quantitatively. This can be done by measurements of the four relevant free energies, that appear in Eq. (3). Kolb and Weber [4] studied the binding equilibria of chicken heart lactate dehydrogenase with oxalate and NADH. Although this enzyme is a tetramer that binds 4 mol of each of the ligands it was possible to demonstrate that in this case the effects at each binding site were wholly independent of the others (see Chapter IV), so that the relations were the same as would be obtained in a protein that bound a single mole of oxalate and NADH. Four dissociation constants were measured: NADH and oxalate in the absence of each other [$K(\text{NADH})$ and $K(\text{OX})$ respectively], oxalate in the presence of saturating concentrations of NADH, $K(\text{OX}/\text{NADH})$, and NADH in the presence of saturating concentrations of oxalate $K(\text{NADH}/\text{OX})$. These constants had the following values:

$$K(\text{NADH}) = 1.7 \pm 0.2 \times 10^{-6}$$

$$K(\text{OX}) = 4.3 \pm 1.5 \times 10^{-3}$$

$$K(\text{NADH}/\text{OX}) = 2.0 \pm 0.3 \times 10^{-7}$$

$$K(\text{OX}/\text{NADH}) = 6.8 \pm 0.4 \times 10^{-4}$$

The two values of δG_{xy} thus determined (-1.3 and 1.1 kcal mol^{-1}) differed by less than 0.3 kcal mol^{-1}, the expected experimental error in the determination of the individual free energies of ligand association.

More recently Reinhart *et al.* have made an important application of Eq. (9) to the analysis of the interactions of substrates and allosteric effectors bound to enzymes [5]. If δG_{xy} is determined at various temperatures a van't Hoff plot permits calculation of the enthalpy change of the disproportionation reaction (8), where X and Y represent the substrate and the allosteric effector, respectively. In agreement with many other cases of ligand–protein equilibria (see Chapter II) the standard enthalpy and entropy contributions are of the same sign and appreciably larger than the standard free energy change of the disproportionation reaction. As a result (Fig. 4) the standard free energy coupling may change rapidly with temperature, disappearing at the temperature of complete compensation ($\Delta H = T \Delta S$) and becoming of opposite sign at temperatures below it. Reinhart *et al.* [5] point out that since the reaction involves the occupation of the same binding sites, whether in the same or different molecules, the contribution of solvent displacement and ligand immobilization to the standard change in entropy must be of very minor importance in relation to that from the degrees of freedom of the protein.

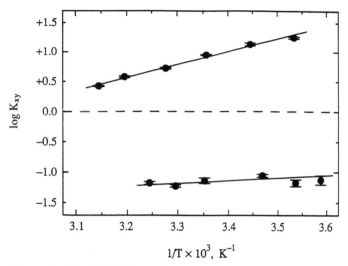

Figure 4. Van't Hoff plot of the free energy of the disproportionation reaction [Eq. (9)] for the binding of two allosteric systems. Upper panel: isocitrate dehydrogenase. Allosteric ligands are isocitrate and ADP. Lower panel: phosphofructokinase. Allosteric ligands are fructose 6-phosphate and ATP. From Reinhart *et al.* [5].

References

1. Weber, G. (1972) *Biochemistry* **11**, 864–873. Weber, G. (1975) *Adv. Protein Chem.* **29**, 1–83.

2. Wyman, J. (1948) *Adv. Protein Chem.* **4**, 410–531. Wyman, J. (1964) *Adv. Protein Chem.* **19**, 224–286. Wyman, J. (1965) *J. Mol. Biol.* **11**, 631–644.

3. Henderson, L.J. (1920) *J. Biol. Chem.* **41**, 401–430. The antagonism of the blood transport of CO_2 and O_2 was previously observed by Bohr, C., Hasselbalch, K., and Krogh, A. (1904) *Scand. Arch. Physiol.* xvi, 2.

4. Kolb, D.A., and Weber, G. (1975) *Biochemistry* **14**, 4471–4476.

5. Reinhart, G.D., Hartleip, S.B., and Symcox, M.M. (1989) *Proc. Natl. Acad. Sci. U.S.A.* **86**, 4032–4036.

IV

Binding of One Ligand at Multiple Protein Sites

Multiple binding is common with proteins. Hemoglobin binds four molecules of oxygen, immunoglobulin G has two equivalent specific binding sites, the NAD-linked dehydrogenases commonly have two or four sites for the binding of coenzyme and an equal number for the binding of substrate, and serum albumin has a few sites at which large hydrophobic anions are bound with high affinity and a much larger number of similar sites of lesser affinity. Proteins undoubtedly offer the best and the most interesting examples of polyvalent molecular complexes.

Thermodynamic Description of Multiple Binding

Let the protein have N sites able to bind a specific ligand. If J out of N such sites are occupied the ligands can be distributed among the binding sites in a number of ways given by the binomial coefficient $\binom{N}{J}$. This is the coefficient of z^J in the binomial expansion of

$$(1 + z)^N = 1 + Nz + \frac{N(N-1)}{2}z^2 + \cdots + z^N \qquad (1)$$

and equals in general

$$\binom{N}{J} = N \cdot (N-1) \cdot (N-2) \cdots \frac{N-J+1}{1 \cdot 2 \cdots J} \qquad (2)$$

Each of the $\binom{N}{J}$ possible combinations of occupied sites in the molecular species PX_J determines a specific site *isomer*, which we designate as

$$PX_J(i), \qquad 1 \le i \le \binom{N}{J} \qquad (3)$$

The standard free energy of formation $PX_J(i)$ from 1 mol of P and J mol of X is

$$\Delta G_{0-J}(i) \tag{4}$$

We recall that the symbol ΔG designates the standard free energy of *association* of the ligand to the protein and that the Ks are the corresponding *dissociation* constants. At equilibrium, two sites, i and k, are occupied by ligand in a ratio determined by their standard free energies of formation:

$$\frac{PX_J(i)}{PX_J(k)} = \exp - \left[\frac{\Delta G_{0-J}(i) - \Delta G_{0-J}(k)}{RT} \right] \tag{5}$$

and the fraction of the PX_J species present as the i site isomer is

$$w_{0-J}(i) = \frac{\exp - \left[\Delta G_{0-J}(i)/RT \right]}{\displaystyle\binom{N}{J} \sum_{i=1} \left[\exp - (\Delta G_{0-J}(i)/RT) \right]} \tag{6}$$

Because of the fixed weight of the site isomers we can assign a definite average free energy of formation $\langle \Delta G_{0-J} \rangle$ to the product PX_J resulting from the reaction

$$P + JX \rightarrow PX_J \tag{7}$$

regardless of the distribution of the Jth ligand among the binding sites. $\langle \Delta G_{0-J} \rangle$ is the standard free energy change that follows the addition of exactly J molecules of ligand to each protein molecule to yield a species in which the site isomers are represented in the proportions indicated by Eq. (6). Following this definition:

$$\langle \Delta G_{0-J} \rangle = \sum_{i=1}^{\binom{N}{J}} w_{0-J}(i)\, \Delta G_{0-J}(i) \tag{8}$$

Moreover we can write

$$\langle \Delta G_{0-J} \rangle = \Delta G_1 + \Delta G_2 + \cdots + \Delta G_J \tag{9}$$

where $\Delta G_1 \ldots \Delta G_N$ represents the average intrinsic free energies of the N

successive additions of ligand X in reactions of the form

$$PX_{J-1} + X \rightarrow PX_J, \qquad 1 \leq J \leq N \tag{10}$$

The corresponding *dissociation* constants are

$$K_J = \exp\left(\frac{\Delta G_J}{RT}\right) \tag{11}$$

To the free energies $\langle \Delta G_{0-J} \rangle$ there correspond products of these dissociation constants:

$$K_1 \cdot K_2 \cdots K_J = \exp\left(\frac{\langle \Delta G_{0-J} \rangle}{RT}\right) \tag{12}$$

The description of multiple equilibria, specifically that of hemoglobin with oxygen, as a series of stepwise equilibria of the form given in Eq. (10) was first proposed by Adair in 1925 [1]. Notice however that equilibria like these may be defined in two ways: either

$$K_J = \frac{[PX_{J-1}][X]}{[PX_J]} \tag{13}$$

or

$$K_J = \frac{(N - J + 1)[PX_{J-1}][X]}{J[PX_J]} \tag{14}$$

The equilibria of Eq. (13) are expressed in terms of macromolecule concentrations and those of Eq. (14) in terms of binding site concentrations. The former are "statistical" dissociation constants and the latter are "intrinsic" dissociation constants. It is these latter constants that are calculated by means of Eq. (11), which defines an average intrinsic free energy of binding. The relation between these two kinds of dissociation constants is given by

$$K_J(\text{stat}) = K_J(\text{intr}) \frac{J}{N - J + 1} \tag{15}$$

Correspondingly

$$\Delta G_J(\text{stat}) = \Delta G_J(\text{intr}) + RT \ln\left(\frac{J}{N - J + 1}\right) \tag{16}$$

The additional term $RT \ln[J/(N - J + 1)]$ is evidently an entropy contribution arising from the multiplicity of binding sites that permits different occupancies for equal number of bound ligands. This entropy contribution is included in the statistical free energy of binding but not in the intrinsic free energy of binding, which solely depends on the average change in chemical potential on binding. The intrinsic constants should be used whenever N is known and in the following text the symbol K will denote them. The difference in numerical value of the two kinds of constants is considerable; for the case of four binding sites:

$$K_4(\text{stat})/K_4 = 1/4$$

$$K_3(\text{stat})/K_3 = 2/3$$

$$K_2(\text{stat})/K_2 = 3/2$$

$$K_1(\text{stat})/K_1 = 4/1$$

Titration of the Protein with Ligand

A titration of P with X is defined by the average number $\langle n \rangle$ of mols of ligand X bound by a mol of protein P as a function of [X], the equilibrium concentration of free ligand. Evidently,

$$\langle n \rangle = \frac{\sum\limits_{J=1}^{N} J[PX_J]}{\sum\limits_{J=1}^{N} [PX_J]} \tag{17}$$

From Eqs. (7), (8), and (12) it follows that

$$[PX_J] = \frac{\binom{N}{J}[P][X]^J}{K_1 \ldots K_J} \tag{18}$$

Introduction of this value in Eq. (17) gives

$$\langle n \rangle = \frac{\sum\limits_{J=0}^{N} J\binom{N}{J}[X]^J/(K_0 \ldots K_J)}{\sum\limits_{J=0}^{N} \binom{N}{J}[X]^J/(K_0 \ldots K_J)} \tag{19}$$

$K_0 = 1$ is added for notational convenience in the last and following equations. This makes

$$\binom{N}{J}[X]^0/K_0 = 1 \tag{20}$$

Equation (19) may be written in a compact form, useful for computational purposes:

$$\sum_{J=0}^{N} a_J[X]^J(\langle n \rangle - J) = 0 \tag{21}$$

where we have set, as we shall usually do in the following, $a_J = \binom{N}{J}/(K_0 \dots K_J)$. Equation (21) will be called the Adair equation and the constants K_1 to K_N the Adair constants. If all N sites have equal affinities for the ligand, $K_1 = K_2 = \cdots K_N = K$. Setting besides $z = [X]/K$,

$$\langle n \rangle = \sum_{J=0}^{N} \frac{\binom{N}{J}Jz^J}{\displaystyle\sum_{J=0}^{N} \binom{N}{J}z^J} \tag{22}$$

The numerator of (22) equals $Nz(1 + z)^{N-1}$ and the denominator equals $(1 + z)^N$ giving

$$\langle n \rangle = \frac{Nz}{1 + z} = \frac{N[X]}{[X] + K} \tag{23}$$

so that in this case the dependence of the fractional occupancy $s = \langle n \rangle/N$ on free ligand concentration is identical to that for a protein with a single binding site. We note that this ready reduction to the form characteristic of equal, independent sites does not occur if, instead of the intrinsic constants, we use the statistical ones and then set these to equal value.

Binding Types

At any point of the titration curve defined by a specified value of the free ligand concentration, an apparent occupancy dependent dissociation constant K_n is defined by the relation

$$K_n = \left[\frac{N - \langle n \rangle}{\langle n \rangle}\right][\text{X}] \qquad (24)$$

A description of the main binding types can be done by reference to the change in K_n with $\langle n \rangle$ or through the properties of appropriate graphic plots of the numerical data. In the Bjerrum plot [2] the logarithm of the free ligand concentration, log[X], is plotted against the average number bound $\langle n \rangle$, or the occupancy with ligand, $s = \langle n \rangle/N$. A complete plot is one in which saturation of P with X is reached at a definite value of $\langle n \rangle$. Saturation is shown by a vertical stretch of the plot, which implies no further change in $\langle n \rangle$ with change in log[X]. An incomplete plot is one in which at the highest values of $\langle n \rangle$ reached there is still a finite change of $\langle n \rangle$ with log[X]. The reason for an incomplete plot is some experimental drawback; the most common is the limited solubility of the ligand. The plot is "resolved" if two or more regions of it are separated by a stretch of higher slope indicating that more than one binding affinity is involved, otherwise it is unresolved. These cases are shown in the schematic plots of Figure 1. If the solubility of the ligand is appropriate the more common case is that of a complete, unresolved plot, and in this case we can distinguish three possibilities [3]:

1. Simple binding characterized by a unique dissociation constant [Eq. (22)], that coincides with the apparent constant [Eq. (24)]. The "characteristic free-ligand span" is defined as the interval of log[X] limited by $\langle n \rangle/N$ values of 0.1 and 0.9. For simple binding this span equals 1.908 decimal logarithmic units.

2. Multiple binding is characterized by an increase in K_n with $\langle n \rangle$ or equivalently $dK_n/d\langle n \rangle > 0$. The Bjerrum plot shows a logarithmic span greater than 1.908 units. It indicates the existence of two or more Adair constants separated by intervals that are not large enough to give rise to a resolved plot. In practice, if $N = 2$, resolution will not be observed unless K_2/K_1 equals 10 to 30, depending upon the precision of the measurements (Fig. 1). Multiple binding will also be observed if there are antagonistic free energy couplings between ligands bound with the same unconditional free energy. This case, which can arise in oligomeric proteins having identical subunits, cannot be formally distinguished from those cases in which there are differences in the unconditional free energies of binding at the different sites.

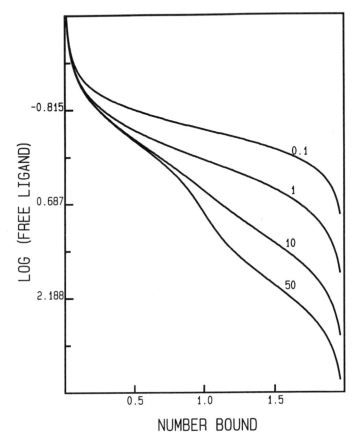

Figure 1. Bjerrum plots for a protein with two binding sites with ratio of the Adair constants K_2/K_1 equal to 0.1 (cooperative binding), 1 (simple binding), 10 and 50 (multiple binding). The first two are unresolved. The plot shown for $K_2/K_1 = 10$ is barely resolved although plotted with increments in $\langle n \rangle$ of 0.01. It would not appear resolved with the lesser density of points of lower precision encountered in practice.

A plot of the logarithmic span against $\log(K_2/K_1)$ has unit slope for values of K_2/K_1 greater than unity (Fig. 2).

3. Cooperative binding is characterized by a decrease in K_n with $\langle n \rangle$, that is $dK_n/d\langle n \rangle > 0$. In a Bjerrum plot it shows a shortened logarithmic span. A decrease in dissociation constant with

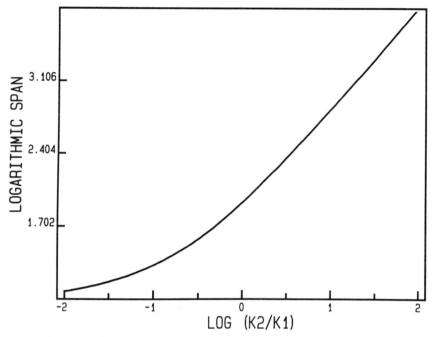

Figure 2. Plot of the logarithmic span of the Bjerrum plot for the range $K_2/K_1 = 0.01$ to $K_2/K_1 = 100$.

number bound can occur only if there is cooperative free energy coupling between the bound ligands.

Apart from the Bjerrum plot two other methods have been employed in distinguishing among these three types of binding.

Klotz method [4]: If s_1 denotes the saturation of ligand by protein then

$$s_1 = \frac{1 - [X]}{X_0} \tag{25}$$

where X_0 is the total ligand concentration. With P_0 as the total protein concentration, the binding sites equal NP_0, the free sites are $NP_0 = s_1 X_0$ and the free ligand is $X_0(1 - s_1)$. The dependence of the apparent dissociation constant on free ligand is given by

$$\langle K \rangle = \frac{(NP_0 - s_1 X_0)(1 - s_1) X_0}{s_1 X_0} \tag{26}$$

or

$$P_0/(s_1 X_0) = \frac{1}{N} + \frac{\langle K \rangle / N}{(1 - s_1) X_0} \qquad (27)$$

According to Eq. (27) a plot of $P_0/(s_1 X_0)$ against $1/[(1 - s_1) X_0]$ has ordinate intercept $1/N$ and Slope $\langle K \rangle / N$. The relation of Klotz plot to that of Bjerrum is made clear by the substitutions:

$$NP_0 = \sum_{J=0}^{N} [PX_J]; \qquad s_1 X_0 = \sum_{J=0}^{N} J[PX_J]; \qquad (1 - s_1) X_0 = [X]$$

which give Eq. (27) the form

$$\frac{1}{\langle n \rangle} = \frac{1}{N}\left(1 + \frac{\langle K \rangle}{[X]}\right) \qquad (27')$$

This latter equation is used in Figure 3: A plot convex toward the abscissa

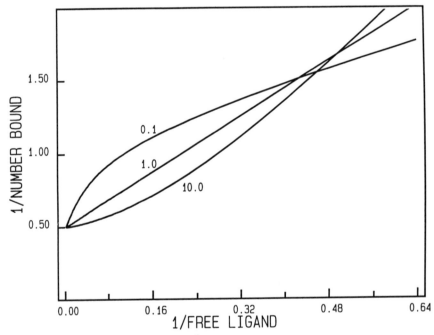

Figure 3. Klotz plot for cases of cooperative binding ($K_1/K_2 = 10$), simple binding ($K_1/K_2 = 1$), and multiple binding ($K_1/K_2 = 1/10$).

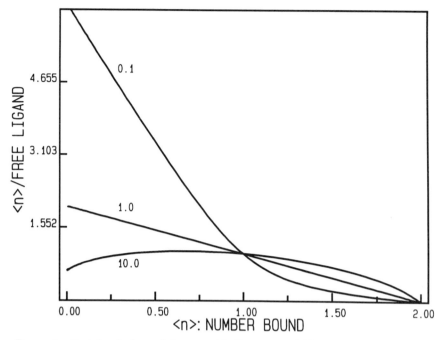

Figure 4. Scatchard plots of the same binding types of Figure 3.

characterizes multiple binding and a concave plot characterizes cooperative binding.

Scatchard method [5]: From Eq. (24)

$$\frac{\langle n \rangle}{[\mathrm{X}]} = \frac{N}{K_n} - \frac{\langle n \rangle}{K_n} \tag{28}$$

A plot of $\langle n \rangle/[\mathrm{X}]$ against $\langle n \rangle$ has intercept N/K_N and differential slope $\langle n \rangle/K_n$. If the plot is convex to the $\langle n \rangle$ axis K_n increases with $\langle n \rangle$, characteristic of multiple binding. If the plot is concave to the $\langle n \rangle$ axis then K_n decreases with $\langle n \rangle$ and the binding is therefore cooperative (Fig. 4).

While the various plots described are in theory equivalent in their ability to distinguish among the binding types, they are in practice of very unequal value. The Bjerrum plot is greatly to be preferred to the other two. In the first place the variables [X] and $\langle n \rangle$ of Eq. (24) are each referred to a separate axis, which renders evident their relation from the graphic. In the plots of Klotz and Scatchard this simplicity is lost. Besides, these two plots permit extrapolation to values of $\langle n \rangle$ that are not experimentally reached, while Bjerrum's logarithmic plot does not allow such extrapolation to be made and therefore obliges the experimentalist to keep within the bounds of his data. In both Scatchard and Klotz methods a single plot is used to extract the number of effective binding sites and the dissociation constants. As commented in Chapter I, these procedures entail a loss of precision in the two determined quantities, and examination of the propagation of the experimental errors to the calculated value of K_n [6] shows that the Bjerrum plot is superior to the other two also in this respect. The Klotz plot according to Eq. (27) has merit when N is unknown and the equilibrium has to be characterized by the effects of binding on a small ligand, the properties of which in the free and bound states are characteristically different.

Independent Binding and Adair Equation

Independent sites are those at which a bound ligand does not show energetic coupling with other ligands. Binding of a ligand X at N independent sites, that may or may not belong to the same protein species, can be described by the equation

$$\langle n \rangle = \sum_{i=a}^{n} \frac{[X]}{[X] + L_i} \tag{29}$$

Here $a, b, \ldots, n$ designate the different binding sites of the molecule or molecules and $L_a, \ldots, L_n$ are their respective dissociation constants. L is used instead of K to sharply emphasize the difference with the Adair constants. L_i depends on the affinity of a specific site while the Adair K_i depends on the affinity of the ith ligand bound to a protein in a series of successive additions. For the case $N = 2$, involving binding at sites a and b Eq. (29) gives,

$$\langle n \rangle = \frac{[X]}{[X] + L_a} + \frac{[X]}{[X] + L_b} \tag{30}$$

or

$$\langle n \rangle = \frac{[X](1/L_a + 1/L_b) + 2[X]^2/(L_a L_b)}{1 + [X](1/L_a + 1/L_b) + [X]^2/(L_a L_b)} \tag{31}$$

This equation adopts the form of an Adair equation by setting

$$\frac{1}{L_a} + \frac{1}{L_b} = \frac{2}{K_1}; \qquad L_a L_b = K_1 K_2 \qquad (32)$$

And from the last two equations

$$K_2 = \frac{L_a + L_b}{2} \qquad (33)$$

Therefore, K_1 and K_2 are, respectively, the harmonic and arithmetic means of L_a and L_b and therefore always either $K_1 = K_2$ or $K_1 < K_2$. It follows that Eq. (29) can never represent a case of cooperative binding. A similar analysis for the case of four bound ligands presents considerable interest, as the literature is rich in errors resulting from the misuse of Eq. (29) for this particular case. Following the procedure adopted for $N = 2$ [Eqs. (30) to (33)] we obtain for this case, which involves sites a, b, c, d:

$$\frac{4}{K_1} = \frac{1}{L_a} + \frac{1}{L_b} + \frac{1}{L_c} + \frac{1}{L_d}$$

$$\frac{6}{K_1 K_2} = \frac{1}{L_a L_b} + \frac{1}{L_a L_c} \cdots + \frac{1}{L_c L_d}$$

$$\frac{4}{K_1 K_2 K_3} = \frac{1}{L_a L_b L_c} \cdots + \frac{1}{L_b L_c L_d}$$

$$K_1 K_2 K_3 K_4 = L_a L_b L_c L_d \qquad (34)$$

From these equations it is relatively straightforward to show that *always*

$$K_1 \le K_2 \le K_3 \le K_4 \qquad (35)$$

In the general case the constants L and K are related by N equations like (34) and the general result is

$$K_1 \le K_2 \le \cdots \le K_N \qquad (36)$$

It is implicit in Eq. (29) that $\langle n \rangle$ is determined by N free energies of formation of *independent* complexes. In consequence the sites are filled in order of their affinity, there being no possibility of describing a binding process of cooperative character, as this is one in which the affinity increases with the number bound. If one attempts to define the free energy change in the reaction $P + JX \rightarrow PX_J$ as done in the Adair formulation, employing Eqs. (34) one obtains for the mean value of the free energy expressions that differ from those generated by means of Eq. (8). Employing the first of Eqs. (34) for $J = 1$ one would thus have

$$\exp\left(\frac{-\Delta G_1}{RT}\right) = \frac{1}{4} \exp\left[\left(\frac{-\Delta G_a}{RT}\right) + \cdots + \exp\left(\frac{-\Delta G_d}{RT}\right)\right] \quad (37)$$

and similar expressions for the equations that yield K_2 to K_4. On comparison with Eq. (8) we note that $\langle \Delta G_{0-J} \rangle$ depends on the distribution of ligands among the sites, which in turn demands specification of the free energy couplings between the ligands bound at the different sites. In Eq. (8) these are implicit in the existence of different values of $\Delta G_{0-J}(i)$ according to such distribution. In contradistinction Eq. (37) calculates the free energy of formation of the PX_J species from the free energies of binding at the separate sites, regardless of the free energy couplings that are set up as the sites are filled. Nevertheless Eq. (37) has been repeatedly used to calculate the free energies of successive additions of oxygen to hemoglobin under the mistaken impression that this equation, which gives the total partition function for a system of *independent* particles, ought to apply equally well when these are interdependent.

The Hill Coefficient

An additional way of distinguishing among the several types of binding is by a numerical parameter, the Hill coefficient [7]. This can be formally defined as the slope in the plot of $\log[(N - \langle n \rangle)/\langle n \rangle]$ against $\log[X]$:

$$H = \frac{d \ln[(N - \langle n \rangle)/\langle n \rangle]}{d \ln[X]} \quad (38)$$

The properties of the system disclosed by this plot are best noticed on examining the case of $N = 2$. For this particular case we set

$$K_1 = Km; \qquad K_2 = K/m \quad (39)$$

so that $K = (K_1 K_2)^{1/2}$ is the geometric mean of the Adair constants. We abbreviate the notation by setting $z = [X]/K$; $R = (N - \langle n \rangle)/\langle n \rangle$. With

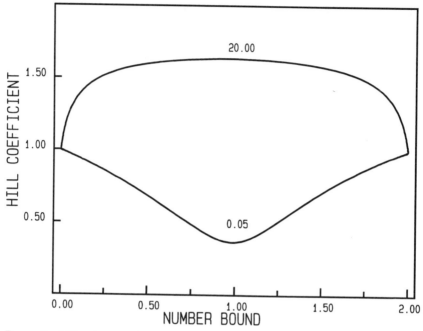

Figure 5. Hill plots to show the variation of H with $\langle n \rangle$ in the cases of cooperative binding of two ligands $(K_2/K_1 = 1/20)$ and multiple binding $(K_2/K_1 = 20)$.

these substitutions Eq. **(19)** becomes

$$R = z \frac{1 + mz}{m + z} \tag{40}$$

$$H = \frac{z}{R} \frac{dR}{dz} = \frac{1 + 2zm + z^2}{1 + z(m + m^{-1}) + z^2} \tag{41}$$

If $m = 1$, then $H = 1$ for any z. If z is either much smaller or much larger than $m + m^{-1}$ then $H \to 1$ for any m (Fig. 5). At midsaturation $(z = 1)$ and

$$H_{1/2} = \frac{2 + 2m}{2 + m + m^{-1}} \tag{42}$$

According to **(42)** if $m < 1$, that is $K_2 > K_1$, $H_{1/2} < 1$ and if $K_2 < K_1$ then $1 < H_{1/2} < 2$. The interaction between the bound ligands is given by

$\delta G_{xx}/RT = -2\ln(m)$. A Hill coefficient above unity always indicates cooperative interactions among bound ligands: If $m = 10$, which implies $K_1/K_2 = 100$, at room temperature $\delta G_{xx} = -2.67$ kcal/mol and $H_{1/2} = 1.82$. A Hill coefficient smaller than unity indicates a heterogeneous population of sites or antagonistic interactions among bound ligands. The Hill coefficient for the general Adair formulation may be derived by proceeding as done in Eqs. (38) to (40). The final result, where we set $F_J = a_J[X]^J$ is [8]

$$H = \frac{\sum_{J=0}^{N} J^2 F_J}{\sum_{J=0}^{N} J F_J} - \frac{\sum_{J=0}^{N} J(N-J)F_J}{\sum_{J=0}^{N} (N-J)F_J} \tag{43}$$

The second ratio of sums in the last equation includes only powers of [X] up to the $(N-1)$th power since $N - J = 0$ for $J = N$. Therefore the maximum value of H is reached when the coefficient a_N greatly exceeds all others and the limiting value is that of the first quotient of sums, which evidently tends to N. When $[X]^J \ll a_J$, that is when $\langle n \rangle \ll 1$ the first quotient tends to 1 and the second, being of order $a_1/(1 + a_1)$ tends to 0. When the dominant term in each sum is $a_N[X]^N$, that is when $\langle n \rangle \to N$, the first quotient of sums tends to N and the second to $N - 1$. Thus H tends to unity as $\langle n \rangle$ tends to either 0 or N and the general case follows the rules obtained for $N = 2$. The Hill coefficient observed at $\langle n \rangle = N/2$ may be taken as an approximate measure of the "order of the binding reaction" that is the effective number of ligand molecules that combine "simultaneously" with the protein, virtually excluding the observation of complexes that involve smaller number of ligands. The Hill coefficient was originally introduced following considerations of this kind [9]: If the equilibrium in which J ligands combine with a protein is formulated as involving no intermediate complexes with less than J ligands the dissociation constant of the PX_J complex is

$$K = \frac{[P][X]^J}{[PX_J]} \tag{44}$$

$$\frac{\langle n \rangle}{J - \langle n \rangle} = \frac{[PX_J]}{[P]} = \frac{[X]^J}{K} \tag{45}$$

and a Hill plot has a coefficient J. According to Eqs. (41) and (43) H is not a constant but a function of $\langle n \rangle$ with the shape shown in Figure 5.

However, in most cases of interest, which are those of cooperative binding, it is reasonably constant for $\langle n \rangle / N$ comprised between $1/4$ and $3/4$. Therefore no gross error is committed by taking $H_{1/2}$, the value at midsaturation, as characteristic for the system.

References

1. The original formulation of stepwise ligand addition is that of Adair, G.S. (1925) *J. Biol. Chem.* **63**, 529–595.

2. Bjerrum, N. (1923) *Z. Phys. Chem.* **106**, 219–242. Bjerrum, J. (1944), who cultivated the paternal plot, *Kgl. Dnsk. Vidensk. Selek. Math, Phys. Med.* **21**(4), 3–7.

3. Weber, G. (1965) In *Molecular Biophysics*, B. Pullman and M. Weissbluth, eds. Academic Press, New York, pp. 369–396.

4. Klotz, I.M., Walker, F.M., and Pivan R.B. (1946) *J. Am. Chem. Soc.* **68**, 1486–1490.

5. Scatchard, G. (1949) *Ann. N.Y. Acad. Sci.* **51**, 660–672. Scatchard, G., Coleman, J.S., and Shen, A.L. (1957) *J. Am. Chem. Soc.* **79**, 12–20.

6. Deranleau, D.A. (1969) *J. Am. Chem. Soc.* **91**, 4044–4049 and 4050–4054. Person, W.B. (1965) *J. Am. Chem. Soc.* **87**, 167–170.

7. Hill, A.V. (1910) Proceedings of the Physiological Society: "The possible effect of the aggregation of hemoglobin on its dissociation curves." *J. Physiol.* **40**, iv–vii.

8. Weber, G. (1975) *Adv. Protein Chem.* **29**, 1–83.

V

Multiple Binding with Ligand Interactions

In some cases, notably in certain oligomeric proteins, the binding sites are known to have structures and properties similar enough to presuppose equal unconditional free energies of binding, but experiment shows that the binding is not of the simple type. The departure from simple binding must arise from interactions between the bound ligands. In the general case each of the N binding sites of the macromolecule may have a different unconditional free energy of binding and from these and the free energy couplings between all the possible pairs of bound ligands one can compute the Adair equation corresponding to the system and compare the predicted titration behavior with the experimental results.

Computation of the Adair Constants

The general procedure consists of the following steps:

1. Determination of a series of Gibbs energy levels $G_0, G_1 \ldots G_N$ corresponding to the average free energy of the protein with $0, 1 \ldots N$ attached ligands, respectively.

2. Computation of the N dissociation constants appearing in the Adair equation by means of N relations of the form

$$\ln K_J = \frac{G_J - G_{J-1}}{RT}, \qquad 1 \leq J \leq N \qquad (1)$$

The disposition of the G levels in a tetramer is shown in Figure 1. As we are interested in the differences between levels and not in their absolute values G_0 can be set equal to zero. Notice also that $G_1 \ldots G_N$ are large negative quantities, since they are generated

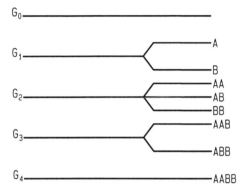

Figure 1. Disposition of the Gibbs energy levels for a molecule with four binding sites, two A sites and two B sites with intrinsic binding free energies $\Delta G^A > \Delta G^B$.

by the association of ligands to the protein at ligand concentrations that are usually very small compared to unity. When J ligands are bound there are $\binom{N}{J}$ possible site isomers, which may differ in free energy. The values of G_J appearing in Eq. **(1)** are averages over the $\binom{N}{J}$ site-isomer levels. As $\binom{N}{0} = \binom{N}{N} = 1$, G_0 and G_N are always unique levels.

3. To determine the N free energy levels $(G_0 \ldots G_N)$ the binding sites are first numbered 1 to N. This number identifies them in the protein and bears no relation to the order in which they are liganded as the concentration of free ligand increases. A value ΔG^i ($i = 1$ to N) is assigned to the *unconditional* free energy of binding at each of the sites.

4. A set of free energy couplings is assigned to each pair of potentially occupied sites, there being $N(N - 1)/2$ such possible pairs. These will be designated δG_{ik} with i and k comprised in the range 1 to N and $i\langle\rangle k$. Note that the lower case indices i, k refer to the individual protein sites, while upper case (capital) subscripts designate as usual the order of binding addition.

The Gibbs free energies of the site isomers of PX_J are $G_J(i \ldots k)$ where $i \ldots k$ are J different numbers comprised between 1 and N that designate the liganded sites. $G_J(i \ldots k)$ is given by the sum of the unconditional free energies of ligand addition at the occupied sites $(\Delta G^j \ldots \Delta G^k)$ and the free energy couplings between them:

$$G_J(i \ldots k) = \Delta G^i + \cdots + \Delta G^k + \sum_{1,2}^{J,J-1} \delta G_{ik} \qquad (2)$$

there being $J(J-1)/2$ terms in the sum of the free energy couplings between the J sites. The weight of this site isomer is given by

$$w_J(i \ldots k) = \frac{\exp - [G_J(i \ldots k)/RT]}{\sum\limits_{1,2}^{J,J+1} \exp - [G_J(i \ldots k)/RT]} \qquad (3)$$

and the average free energy G_J of the PX_J species is

$$G_J = \sum_{1,2}^{J,J-1} G_J(i \ldots k) w_J(i \ldots k) \qquad (4)$$

An Example: Adair Constants of a Molecule with Four Coupled Binding Sites

As an example we detail the calculation of G_2 for the tetramer molecule with two α and two β binding sites, each kind having a fixed unconditional free energy of binding. These and the free energy couplings between pairs of bound ligands are in RT units:

$$\Delta G^1 = \Delta G^3 = 0$$

$$\Delta G^2 = \Delta G^4 = 1$$

$$\delta G_{12} = \delta G_{34} = 0$$

$$\delta G_{14} = \delta G_{23} = -2$$

$$\delta G_{13} = 0$$

$$\delta G_{24} = -1$$

The couplings are graphically shown in Figure 2. For $J = 2$ the combinations of pairs of occupied sites are $\binom{4}{2} = 6$. Application of Eq. (2) yields

$$G_2(1,2) = G_2(2,4) = G_2(3,4) = 1; \qquad G_2(1,4) = G_2(2,3) = -1$$

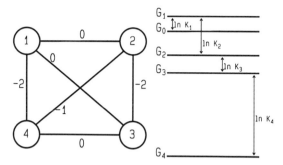

Figure 2. Coupling scheme for four binding sites discussed in text. Coupling free energies between sites, in *RT* units, are shown aside the lines joining the sites. The resulting Gibbs free energy levels and the dissociation constants are shown at the right.

$G_2(1, 3) = 0$, and Eq. (**3**) gives:

$$w_2(1, 4) = w_2(2, 3) = 0.36051$$

$$w_2(1, 2) = w_2(2, 4) = w_2(3, 4) = 0.04879$$

$$w_2(1, 3) = 0.13262$$

Finally, by application of Eq. (**4**), $G_2 = -0.57465$. In similar fashion we determine $G_1 = 0.26894$, $G_3 = -1$, and $G_4 = -3$. The Adair constants for this case are by Eq. (**1**): $K_1 = 1.309$, $K_2 = 0.430$, $K_3 = 0.653$, and $K_4 = 0.135$. They stand in the ratios 9.70:3.18:4.83:1.

Multiple Binding Involving Ligands of Two Types

The simplest type of this kind is one in which 2 mol of a ligand X and 1 mol of ligand Y are involved, and there are energetic couplings between ligands of different kind, but not among the two X ligands. For simplicity we assume that the two X–Y free energy couplings have identical values:

$$X - \delta G_{xy} - Y - \delta G_{xy} - X$$

The equilibria involved and the dissociation constants of the complexes are shown in the accompanying scheme.

$$
\begin{array}{ccc}
Y + P & \overset{K(Y)}{\longleftrightarrow} & YP \\
K(X)\Big| & & \Big| K(X/Y) \\
PX & & YPX \\
K(X)\Big| & & \Big| K(X/Y) \\
PX_2 & & YPX_2
\end{array}
$$

The average number of moles of X bound is

$$
\langle n_x \rangle = \frac{[PX] + [YPX] + 2[PX_2] + 2[YPX_2]}{2\{[P] + [PX] + [YPX] + [PX_2] + [YPX_2]\}} \tag{5}
$$

If [Y], the concentration of free Y, is fixed the titration curve is a function of the parameters $\epsilon = [Y]/K(Y)$ and $b = K(X/Y)/K(X)$. Referring all concentrations appearing in (5) to that of unliganded protein, [P] we have

$$
\frac{[PX]}{[P]} = \frac{2[X]}{K(X)}; \qquad \frac{[PX_2]}{[P]} = \left(\frac{[X]}{K(X)} \right)^2;
$$

$$
\frac{[YP]}{[P]} = \frac{[Y]}{K(Y)} = \epsilon; \qquad \frac{[YPX]}{[P]} = \left(\frac{2[X]}{K(X)} \right) \epsilon b^{-1};
$$

$$
\frac{[YPX_2]}{[P]} = \left(\frac{[X]}{K(X)} \right)^2 \epsilon b^{-2} \tag{6}
$$

From (4) and (5)

$$
\frac{\langle n_x \rangle}{2} = \frac{(1 + \epsilon b^{-1})([X]/K(X)) + (1 + \epsilon b^{-2})([X]/K(X))^2}{1 + \epsilon + (1 + \epsilon b^{-1})(2[X]/K(X)) + (1 + \epsilon b^{-2})([X]/K(X))^2} \tag{7}
$$

Dividing numerator and denominator by $1 + \epsilon$ and setting

$$K_1 = K(X)b\frac{(1 + \epsilon)}{(b + \epsilon)} \tag{8}$$

$$K_2 = K(X)\frac{b(b + \epsilon)}{(b^2 + \epsilon)} \tag{9}$$

Eq. (7) takes the familiar Adair form. From Eqs. (8) and (9) we obtain

$$\frac{K_1 - K_2}{K(X)} = b\epsilon\frac{(b - 1)^2}{(b + \epsilon)(b^2 + \epsilon)} \tag{10}$$

The right-hand side is always positive, whether b be smaller or greater than 1, so that always $K_1 \geq K_2$ and the X–Y interactions result in the cooperative binding of X whether $\delta G_{xy} > 0$ or $\delta G_{xy} < 0$. It is further evident that $K_1 - K_2 \to 0$ when $\epsilon \to 0$ or when [Y] is large enough to make ϵb^{-1} and ϵb^{-2} large compared to unity. At these two extremes of Y concentration the cooperativity in the binding of X disappears. When the right-hand side of Eq. (10) is differentiated with respect to the parameter ϵ and the result set to zero we find as the condition for maximum value of $K_1 - K_2$:

$$\epsilon^2 = b^3 \tag{11}$$

and introducing this condition in Eqs. (8) and (9)

$$K_1/K_2 = \frac{1 + b^{3/2}}{b^{1/2}(1 + b^{1/2})} \tag{12}$$

The ratio K_1/K_2 of Eq. (12) is unchanged if b is replaced by $1/b$, showing that the maximum cooperativity is the same for a given value of δG_{xy}, whether this is positive or negative. The titration of protein with Y at a fixed concentration of X, is derived in a fashion analogous to that of Eq. (7) to give

$$\langle n_y \rangle = \frac{[Y]}{[Y] + K(Y)\big[([X] + K(X))/([X]/b) + K(X))^2\big]} \tag{13}$$

The effective dissociation constant of the Y-protein complex is a function of [X]. For $[X]/b \ll K(X)$,

$$\langle n_y \rangle \to \frac{[Y]}{[Y] + K(Y)} \tag{14}$$

and for $[X]/b \gg K(X)$ the dissociation constant for Y goes into $K(Y)b^2$.

The preceding treatment gives a formal explanation of the cooperativity of X binding induced by the addition of Y irrespective of the character, positive or negative, of the free energy coupling between the two ligands, but the origin of this effect can be appreciated intuitively: In the case of antagonism between X and Y we may fix the concentration of Y so that there is almost, though not complete, saturation of its site in the absence of X. Such Y concentration will be one intermediate between $K(Y)$ and $K(Y/X)$. On increase of X no appreciable addition of this ligand will result until [X] is close to $K(X/Y)/10$ and as X is bound it will result in a release of Y since its free concentration is less than $K(Y/X)$. The second site for X will be now more readily filled and K_2 will roughly correspond to $K(X)$ rather than $K(X/Y)$ as was the case with the first X-ligand bound. The cooperativity resulting when $dG_{xy} < 0$ is similarly explained. In this case $K(Y/X) < K(Y)$ and a concentration of Y intermediate between these two values will be insufficient to produce substantial Y-binding in the absence of X. On [X] approaching $K(X/Y)$ simultaneous binding of X and Y ligands will readily result, and will be followed by the filling of the second X-site since its dissociation constant has changed from $K(X)$ to the smaller value $K(X/Y)$. It follows from this description that cooperativity due to the presence of Y disappears if its concentration is appreciably greater than $K(Y/X)$ or $K(X)$, whichever is larger. In this case the site for Y will always be occupied and the affinity for X will remain constant at all values of $\langle n_x \rangle$. A number of examples in the literature are readily explained on these principles. Thus Tyuma and collaborators [1] found that the addition of 2,3-diphosphoglycerate (DPG), which is antagonistic to oxygen in the hemoglobin–O_2 equilibrium (see Chapter VIII), produces an enhancement of the cooperativity as shown by an increase in Hill coefficient from 2.6 to a value somewhat greater than 3. Such an effect must be dependent on the ratio of DPG concentration to the dissociation constant of the DPG–hemoglobin complex, and occurs following the release of DPG from the complex during oxygenation. Replacement of DPG by a similar concentration of the much more strongly bound inositol hexaphosphate (IHP) [2] results, as also observed by Tyuma *et al.* [2], in an even larger decrease in the affinity of hemoglobin for oxygen but with no increase in cooperativity. The larger affinity of IHP prevents its release during oxygenation. These differences between the actions of DPG and IHP find a natural explanation in terms of the preceding description of coupling effects.

A demonstration of the appearance of cooperativity by addition of a second ligand when there was none in its absence has been given by Kolb and Weber [3]. Since the original studies of Klotz [4] and others [5] it has been known that serum albumin can bind 2–5 mol of a variety of organic anions with high affinity and a considerably larger number with much less affinity. Pasby [6] (1969) showed that four molecules of 1-anilinonaphthalene 8-sulfonate (ANS) are bound at sites of high affinity with dissociation constant nearly micromolar. Binding at such sites [7] is characterized by enhancement of the fluorescence over that of the free ligand in water by a factor of over 100. A much larger number of ANS molecules are bound with lesser affinity and very small increase in fluorescence efficiency. The existence of these two classes of sites is evident in Figure 3, which presents a Bjerrum plot of the binding of ANS by albumin measured by fluorescence enhancement and by equilibrium dialysis. The fluorescence enhancement data define a complete titration curve while the equilibrium dialysis experiment yields an incomplete plot. It was concluded that in the strong binding sites ANS is protected form the quenching action of water and fluoresces with high yield while in the weaker sites it must be in more direct contact with water and is thus weakly fluorescent. An anion more hydrophylic than ANS, like dihydroxybenzoate, would occupy preferentially the hydrophyilic sites and should exercise an antagonistic effect—probably by electrostatic repulsion—on the binding of ANS to the hydrophobic sites. Figure 4 shows clearly the existence of cooperative binding of ANS at intermediate concentrations of dihydroxybenzoate. A Hill coefficient of 1.5 is reached when three ANS molecules are bound in the presence of intermediate concentrations of dihydroxybenzoate. At larger concentrations of dihydroxybenzoate ANS binding regains its simple (noncooperative) binding character.

Polyvalent Interactions of Bound Ligands

We limit our analysis to the case in which the protein concentration is sufficiently small in comparison with the total concentration of the effector ligand Y so that this concentration is not appreciable modified by the uptake or release of Y from the protein complexes. This supposition is not very restrictive as it applies to most cases of interest. It simplifies considerably the analysis and permits appreciation of the relations between the binding of the ligands and the parameters introduced. The more general case in which no restrictions are imposed on the concentrations of ligands or effectors can be solved by numerical computation [8].

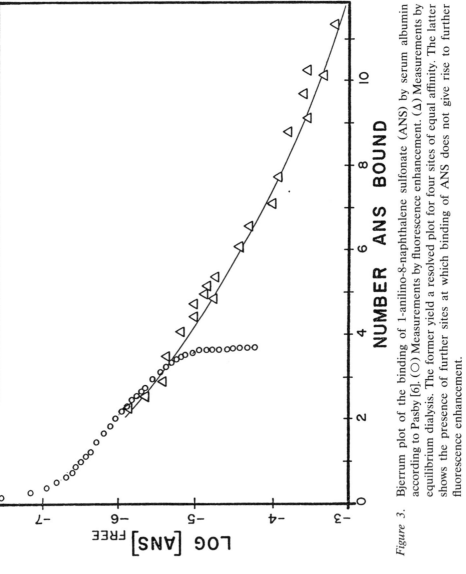

Figure 3. Bjerrum plot of the binding of 1-anilino-8-naphthalene sulfonate (ANS) by serum albumin according to Pasby [6]. (O) Measurements by fluorescence enhancement. (Δ) Measurements by equilibrium dialysis. The former yield a resolved plot for four sites of equal affinity. The latter shows the presence of further sites at which binding of ANS does not give rise to further fluorescence enhancement.

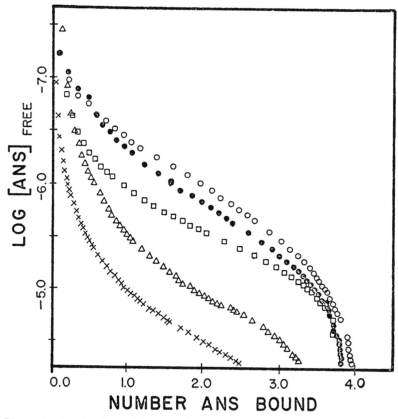

Figure 4. The binding of ANS by albumin in the absence of competitor and in the presence of, right to left, 0, 0.2 mM, 2 mM, 20 mM and 60 mM 3,5-dihydroxybenzoate. Note the appearance of cooperativity at the lower concentrations of dihydroxybenzoate and the disappearance at the highest one. From ref. [3].

The possible polyvalent complexes $Y_K PX_J$ result form K additions of Y and J additions of X to a protein that has N binding sites for X and M binding sites for Y. The average numbers of X bound per protein molecule at [Y] = constant is

$$\langle n_x \rangle = \frac{\displaystyle\sum_{K=0}^{M} \sum_{J=0}^{N} J[Y_K PX_J]}{\displaystyle\sum_{K=0}^{M} \sum_{J=0}^{N} [Y_K PX_J]} \tag{15}$$

writing it as an explicit sum of the concentrations of species with the same number, K of Y ligands,

$$\langle n_{x'K} \rangle = \frac{\sum\limits_{K=0}^{M} [Y_K PX] + 2[Y_K PX_2] + \cdots + N[Y_K PX_N]}{\sum\limits_{K=0}^{M} [Y_K PX] + [Y_K PX_2] + \cdots + [Y_K PX_N]} \tag{16}$$

Moreover, with [Y] constant we can set

$$[Y_K PX_J] = \binom{M}{K}\binom{N}{J}[P]\left(\frac{[X]^J}{K_1(X/Y_K)\ldots K_J(X/Y_K)} \right)$$

$$\times \left(\frac{[Y]^K}{K_1(Y)\ldots K_K(Y)} \right) \tag{17}$$

where the Adair constants $K_J(X/Y_K)$ for the binding of the Jth X-ligand are conditional dissociation constants depending on the previous binding of K Y-ligands. The ratio of products

$$K_1(X/Y_K)\ldots K_J(X/Y_K)/[K_1(X)\ldots K_J(X)]$$

is determined by the *total* free energy couplings $\delta G_{xy}(JK)$ between the X and Y ligands in the $Y_K PX_J$ complex:

$$\frac{K_1(X/Y_K) \cdots K_J(X/Y_K)}{K_1(X) \cdots K_J(X)} = b_{JK} \tag{18}$$

The free energy couplings determine also the reciprocal condition

$$\frac{K_1(Y/X_J)\ldots K_K(Y/X_J)}{K_1(Y)\ldots K_K(Y)} = \exp\left(\frac{\delta G_{xy}(JK)}{RT} \right) = b_{JK} \tag{18'}$$

It follows from Eq. (18) that the conditional Adair dissociation constants for X in this complex are related to the unconditional ones by the equations

$$K_J(X/Y_K) = K_J(X)\frac{b_{JK}}{b_{J-1,K}} \tag{19}$$

and from Eq. (18′) the analogous dissociation constants for Y relate to the unconditional ones as

$$K_K(Y/X_J) = K_K(Y)\frac{b_{JK}}{b_{J,K-1}} \qquad (19')$$

We note that $b_{0K} = b_{J0} = 1$. Therefore

$$b_{11} = \frac{K_1(X/Y)}{K_1(X)}$$

so that from this last relation and Eq. (19) we can derive by recursion the coefficients b_{JK} from the conditional Adair dissociation constants. Thus the description of the binding of two interacting ligands that can occupy, respectively, M and N sites involves $M \times N$ interaction factors, or a corresponding number of conditional Adair dissociation constants, besides the Adair constants for the binding of each separate ligand. Calculation of the total free energy couplings in the $Y_K PX_J$ complex is done as in the case of a single ligand: The number of site isomers in this complex is $\binom{M}{K}\binom{N}{J}$. The free energy couplings in the site isomer that involve the lth isomer of PX_J and the mth isomer of $Y_K P$ will be denoted by $\delta G_{xy}(JK)_{lm}$ with l comprised between 1 and $\binom{N}{J}$ and m comprised between 1 and $\binom{M}{K}$. The weight of this isomer in the equilibrium mixture is

$$w_{lm} = \frac{\exp - \{\delta G_{xy}(JK)_{lm}/RT\}}{\displaystyle\sum_{l=1,\,m=1}^{\binom{N}{J}\binom{M}{K}} \exp - (\delta G_{xy}(JK)_{lm}/RT)} \qquad (20)$$

Therefore

$$\delta G_{xy}(JK) = \sum_{l=1,\,m=1}^{\binom{N}{J}\binom{M}{K}} \delta G_{xy}(JK)_{lm} w_{lm} \qquad (21)$$

Equation (17) may now be written in terms of the free energy couplings $dG_{xy}(JK)$, or the b_{JK} coefficients: To this purpose we define $M + 1$ polynomials in $[Y_0]$, the concentration of Y held constant during the titration with X to yield the species $P(Y/X_J)$, with $J = 0$ to M.

$$P(Y) = 1 + \binom{M}{1}\frac{[Y_0]}{K_1(Y)} + \cdots + \binom{M}{M}\frac{[Y_0]^M}{K_1(Y)\ldots K_M(Y)}$$

$$P(Y/X_J) = 1 + \binom{M}{1}b_{1J}^{-1}\frac{[Y_0]}{K_1(Y)} + \cdots$$

$$+ \binom{M}{M}b_{MJ}^{-1}\frac{[Y_0]^M}{K_1(Y)\ldots K_M(Y)}, \qquad 1 \le J \le N \qquad (22)$$

$K(Y)_1 \ldots K(Y)_M$ being the M Adair constants for the independent binding of Y. From this last equation and (17)

$$\langle n_x \rangle = \frac{\displaystyle\sum_{J=0}^{N} JF_J([X])P(Y/X_J)}{\displaystyle\sum_{J=0}^{N} F_J([X])P(Y/X_J)} \qquad (23)$$

with

$$F_J([X]) = \frac{\binom{N}{J}[X]^J}{K_1(X)\ldots K_J(X)} \qquad (24)$$

We observe that the binding of X in the presence of a constant concentration of Y is described by N Adair constants, which are now explicit functions of $[Y] = [Y_0]$. We draw attention to this dependence by designating them as $K_J(X/[Y])$:

$$K_1(X/[Y]) = K_1\frac{P(Y)}{P(Y/X)}$$

$$K_2(X/[Y]) = K_2\frac{P(Y/X)}{P(Y/X_2)}$$

$$K_J(X/[Y]) = K_J\frac{P(Y/X_{J-1})}{P(Y/X_J)} \qquad (25)$$

and

$$\langle n_x \rangle = \frac{\sum_{J=0}^{N} J[X]^J/(K_1(X/[Y])\dots K_J(X/[Y]))}{\sum_{J=0}^{N} [X]^J/(K_1(X/[Y])\dots K_J(X/[Y]))} \qquad (26)$$

For the case $M = 1$, $N = 2$ Eqs. (25) and (26) reduce to Eqs. (8) or (9) and (7), respectively.

Multiple X-Y Couplings for the Case $N = 4$

We examine now in more detail the case in which the binding of a single effector ligand is energetically coupled to four ligands of a different kind. Setting $M = 1$, $N = 4$ Eqs. (24) give

$$P(Y) = 1 + \frac{[Y]}{K_1(Y)} = 1 + \epsilon$$

$$P(Y/X_J) = 1 + \left(\frac{[Y]}{K_1(Y)}\right)b_{1J}^{-1} = 1 + \epsilon b_{1J}^{-1}, \qquad 1 \le J \le 4 \qquad (27)$$

Therefore

$$K_1(X/[Y]) = K_1(X)b_{11}(1 + \epsilon)/(b_{11} + \epsilon)$$

$$K_2(X/[Y]) = K_2(X)(b_{21}/b_{11})(b_{11} + \epsilon)/(b_{21} + \epsilon)$$

$$K_3(X/[Y]) = K_3(X)(b_{31}/b_{21})(b_{21} + \epsilon)/(b_{31} + \epsilon)$$

$$K_4(X/[Y]) = K_4(X)(b_{41}/b_{31})(b_{31} + \epsilon)/(b_{41} + \epsilon) \qquad (28)$$

We note that if $\epsilon \to 0$ then $K_J(X/[Y]) \to K_J(X)$ and if $\epsilon \gg b_{J1}$ then $K_J(X/[Y]) \to K_J(X_J/Y)$ in agreement with Eq. (19).

If the free energy couplings of all possible ligand associations of X and Y are specified, together with the effector-free Adair constants $K_J(X)$, the corresponding Adair constants for increasing number of bound effector molecules, $K_J(X/Y_K)$ can be derived as described above. With these values and the concentration of effector the titration curve of the protein with X is calculated. We shall find such calculation useful: The free-energy coupling of a single effector molecule with others that are multiply bind plays an important role in the physiological regulation at the molecular level of the oxygen–hemoglobin equilibrium (Chapter VIII) and possibly in some other equilibria.

References

1. Tyuma, I., Schimidzu, K., and Imai, K. (1971) *Biochem. Biophys Res. Commun.* **43**, 423–428.
2. Tyuma, I., Schimidzu, K., and Imai, K. (1971) *Biochem. Biophys Res. Commun.* **44**, 682–686.
3. Kolb, D.A., and Weber, G. (1975) *Biochemistry* **14**, 4475–4481.
4. Klotz, I.M., Walker, F.M., and Pivan, R.B. (1946) *J. Am. Chem. Soc.* **68**, 1486–1490. Klotz, I.M. (1946) *Arch. Biochem.* **9**, 109–117.
5. Kharush, F.J. (1950) *J. Am. Chem. Soc.* **72**, 2705–2713. Kharush, F. (1954) *J. Am. Chem. Soc.* **76**, 5536–5542.
6. Pasby, T. (1969) Doctoral Thesis, University of Illinois.
7. Laurence, D.J.R. (1952) *Biochem. J.* **51**, 168–179. Laurence, D.J.R. (1972) In *Physical Methods in Macromolecular Chemistry*, Vol. 2. B. Carroll, ed. Dekker, New York, pp. 91–183.
8. Royer, C.A., Smith, W.R., and Beecham, J.M. (1990) *Anal. Biochem.* **191**, 287–294.

VI

Ligand Distribution

Ligand Distributions in Simple Binding

The distribution of ligands among the protein molecules depends on the unconditional free energies of binding and the possible interligand interactions. An example of the variation of the distribution of two ligands of different kind, X and Y, with changes in the free energy coupling between them has already been discussed in Chapter III. We derive here the distribution of ligands of a single kind when its binding is described by N Adair constants [1]. The fraction of protein molecules that have J ligands bound is

$$f_J = \frac{[PX_J]}{\sum\limits_{J=0}^{N} [PX_J]} \tag{1}$$

or more explicitly

$$f_J = \frac{a_J[X]^J}{\sum\limits_{J=0}^{N} a_J[X]^J} \tag{2}$$

It follows that the distribution $f_1 \ldots f_N$ can always be calculated if the Adair constants are given. In the case of independent sites of equal ligand affinities all the Ks are equal. Setting then $z = [X]/K$ [Eq. (IV.22)],

$$f_J = \binom{N}{J} \frac{z^J}{(1+z)^N} = \binom{N}{J} \left[\frac{z}{(1+z)} \right]^J \frac{1}{(1+z)^{N-J}} \tag{3}$$

f_J is therefore the Jth term of the binomial expansion of $1/(1 + z)^N$, and represents the probability of occurrence of J events in N trials when the probability of an isolated event is $z/(1 + z)$. This quantity can be recognized as the a priori probability of binding

$$p = [X]/([X] + K) \tag{4}$$

while $1/(1 + z) = K/([X] + K)$ is the a priori probability that binding will not take place. As $z/(1 + z) = \langle n \rangle/N$, Eq. (3) may be written as a function of average number bound $\langle n \rangle$ instead of free ligand z:

$$f_J = \binom{N}{J}\left(\frac{\langle n \rangle}{N}\right)^J\left[1 - \left(\frac{\langle n \rangle}{N}\right)\right]^{N-J} \tag{5}$$

By setting $df_J/d\langle n \rangle = 0$ in Eq. (5) it is found that the maximum of f_J is found at

$$\langle n \rangle = J \tag{6}$$

This latter condition is also valid for arbitrary values of the Adair constants, and therefore it applies in the case of multiple binding with or without ligand interactions. To demonstrate this property we set

$$\frac{df_J}{d\langle n \rangle} = \frac{df_J/d[X]}{d\langle n \rangle/d[X]} \tag{7}$$

and from Eq. (2)

$$df_J/d[X] = \frac{a_J[X]^{J-1}(\langle n \rangle - J) \sum\limits_{J=0}^{N} a_J[X]^J}{\left(\sum\limits_{J=0}^{N} a_J[X]^J\right)^2} \tag{8}$$

since $d\langle n \rangle/d[X] > 0$ for any $\langle n \rangle$ the condition of maximum is that $df_J/d[X] = 0$, which by the last equation is $\langle n \rangle = J$.

Ligand Distributions for Arbitrary Adair Constants

We start by examining the simplest case, $N = 2$. Expressing the two Adair constants as the ratio of each of their geometric mean K we have

$$K_1 = mK; \qquad K_2 = K/m \tag{9}$$

where $m = \sqrt{(K_1/K_2)}$ is an arbitrary number. Setting $z = [X]/K$ [Eq. (IV.39)]

$$\langle n \rangle = \frac{2\left[(z/m) + z^2\right]}{1 + 2(z/m) + z^2} \tag{10}$$

At $z = 1$, $\langle n \rangle = 1$: In the titration of a system with two binding sites the point of midsaturation determines the geometric mean of the Adair constants. At $\langle n \rangle = 1$ application of Eq. (1) gives

$$f_0 = f_2 = \frac{m/2}{1 + m}, \qquad f_1 = \frac{1}{1 + m} \tag{11}$$

The three main binding types have characteristic distributions at $\langle n \rangle = 1$. If $m = 1$, the binding sites have equal affinities for X and

$$f_0 = f_2 = 1/4; \qquad f_1 = 1/2 \tag{12}$$

corresponding to the random distribution of ligands among the protein molecules. As m increases above 1, f_1 increases and f_0 and f_2 decrease, and for $m \gg 1$ f_1 tends to 1. In this case a Bjerrum plot is well resolved and at $\langle n \rangle = 1$ the protein population is virtually homogeneous comprising only PX complexes. As m falls below unity f_1 decreases with respect to f_0 and f_2 and tends to 0 when $m \ll 1$. In this limiting case the cooperative binding of the ligands eliminates the intermediate form PX, and the population consists of equal amounts of f_0 and f_2. The distributions for these different cases in the interval in $\langle n \rangle$ of 0 to 2 are shown in Figure 1.

The correspondence between ligand distribution types and binding types should lead us to inquire on the advantages of determining the latter by a study of the former property. If the ligand distribution were obtained by a method able to distinguish according to number bound, rather than inferred from the titration of protein with ligand, one could apply it at protein concentrations at which free ligand would be too small to be precisely determined, that is under stoichiometric binding conditions. Such direct methods would be particularly useful as it is often the case that the precision of the titration data is hampered by the high dilutions at which the measurements have to be performed to establish an observable equilibrium between protein and ligand. However, there are few such direct methods: Thus Chien Ho and collaborators [1] have shown that some properties of the distribution of the bound dioxygen among the hemoglobin molecules can be obtained by NMR methods, as it is possible to distinguish between liganded α and β hemoglobin chains. This method could be

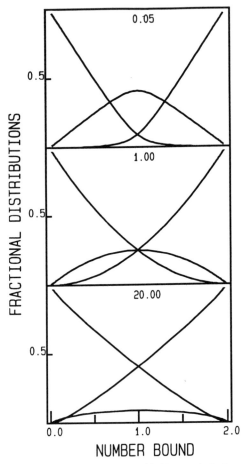

Figure 1. Distributions $f(\langle n \rangle)$ vs. $\langle n \rangle$ for
$N = 2$. Middle panel: Normal dis-
tribution $(K_1 = K_2)$. Upper panel:
Case in which $K_1 = K_2/20$, corresponding to two independent
sites with different ligand affinity.
Lower panel: Cooperative case
$(K_1 = 20K_2)$. Notice the symmetry of the distributions about $N/2$
for all three cases.

similarly successful in other cases of multimer proteins with distinguishable subunits but holds less promise for those cases in which the subunits are identical. A further method of limited application makes use of the possible electronic energy transfer among bound fluorescent ligands. The distances between the binding sites of single-chain and multimer proteins (5 to 30 Å units) are of the same order as those at which electronic energy transfer among identical fluorophores can take place with good efficiency. Such transfer results in a change in direction of the oscillator responsible for the emission of radiation, and as a result the number of ligands bound to the individual molecules determines the polarization of the emitted radiation. The polarization is a maximum for molecules binding a single ligand and decreases monotonically with the number of ligands bound. In the case of a protein with two binding sites of equal free energy of binding the decrease in anisotropy of emission (see Chapter XII) as a function of number bound takes a particularly simple form. If the fluorescence emitted by a pure single-liganded population has a characteristic anisotropy r_1 and that from a doubly liganded population has anisotropy r_2, then a mixed population in which these species are present as fractions f_1 and f_2 emits fluorescence with anisotropy

$$\langle r \rangle = \frac{r_1 f_1 + 2 r_2 f_2}{f_1 + 2 f_2} \tag{13}$$

Expressing the fractions f_1 and f_2 as functions of $\langle n \rangle$ by means of Eq. (5) gives

$$\langle r \rangle = r_1 (2 - \langle n \rangle) + r_2 \langle n \rangle \tag{14}$$

Thus for this case a plot of $\langle r \rangle$ against $\langle n \rangle$ is a straight line which extrapolates to, respectively, r_1 and r_2 at $\langle n \rangle = 0$ and $\langle n \rangle = 2$. At any intermediate value of $\langle n \rangle$ the polarization determines the proportions of singly and doubly liganded molecules. If $K_1 < K_2$ (multiple binding) the plot lies above this line and it lies below it if the binding is cooperative, that is $K_1 > K_2$. For a protein with two binding sites this method is capable of yielding exactly the same information regarding the ligand distribution, and the ratio of the dissociation constants, as the information obtained from the titration data. When N is greater than 2 fluorescence depolarization, or quenching, by energy transfer can still give very valuable information on the ligand distribution, without, however, matching completely the information from titration data, as Weber and Daniel [9] showed in a study of the binding of 1-anilino-8-naphthalene sulfonate by serum albumin (Fig. 2). Lehrer and Leavis [3] have used similar concepts

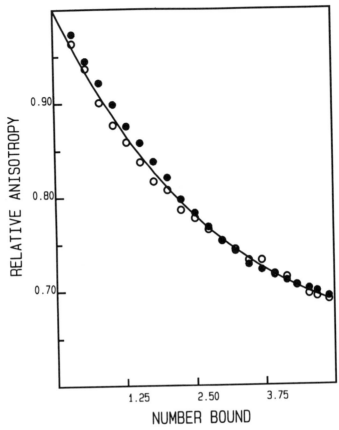

Figure 2. Anisotropy of emission of 1-anilino-8-naphthalene sulfonate bound to serum albumin plotted against average number bound. White circles at pH 7, dark circles at pH 5. From Weber and Daniel [2].

in a study of the quenching of the fluorescence that follows transfer of electronic energy from a fluorescent chromophore to a quencher ligand.

Symmetry of Ligand Distribution and of Titration Data

The normal or Bernoulli distribution of ligands [Eq. (5)] is a symmetric distribution in the sense that

$$f_J(\langle n \rangle) = f_{N-J}(N - \langle n \rangle) \tag{15}$$

$f_J(\langle n \rangle)$ designates here the fraction with J ligands when an average of n ligands per protein are bound. Relation (15) follows from (5) since

$$\binom{N}{J} = \binom{N}{N-J} \tag{16}$$

To determine the condition of symmetry (13) for arbitrary ligand distributions described by Adair equations we define two polynomials:

$$POL_{\langle n \rangle} = \sum_{J=0}^{N} a_J[X]_{\langle n \rangle}^J \tag{17}$$

where $[X]_{\langle n \rangle}$ is the free ligand concentration at $\langle n \rangle$ ligands bound. Similarly

$$POL_{N-\langle n \rangle} = \sum_{J=0}^{N} a_J[X]_{N-\langle n \rangle}^J \tag{18}$$

From the last two equations, after introduction of (13)

$$\frac{POL_{\langle n \rangle}}{POL_{N-\langle n \rangle}} = \frac{K_0 \dots K_{N-J}[X]_{\langle n \rangle}^J}{K_0 \dots K_J[X]_{N-\langle n \rangle}^{N-J}} \tag{19}$$

Equation (17) is valid for all J. From the values for J and $J+1$, respectively, we obtain

$$[X]_{\langle n \rangle}[X]_{N-\langle n \rangle} = K_{J+1}K_{N-J} \tag{20}$$

According to the last equation the constancy of the product $[X]_{\langle n \rangle}[X]_{N-\langle n \rangle}$ provides a simple quantitative criterion to demonstrate the symmetry of ligand distribution [4]. The product constancy shows in a Bjerrum plot as a symmetry about the midpoint, $N/2$. For $N = 2$, setting successively $J = 0$ and $J = 1$ in Eq. (20) we obtain

$$K_1 K_2 = K_1 K_2 \tag{21a}$$

from which it may be deduced that in this case the distribution is always symmetric. This property can be demonstrated independently: for example, we showed above that $f_0 = f_2$ for any values of the dissociation constants. In a similar way we obtain as the condition of symmetry for $N = 3$,

$$K_1 K_3 = K_2^2$$

and for $N = 4$,

$$K_1 K_4 = K_2 K_3 \qquad \text{(21b)}$$

If the constants are expressed in terms of the Gibbs free energy levels G_J for the stepwise addition of ligands [Eq. (V.1)] the last equation becomes, on setting $G_0 = 0$

$$2G_1 + G_4 = 2G_3 \qquad \text{(22)}$$

independent therefore of G_2. If the unconditional free energies of binding are all equal any symmetry of the distribution must be due to the interligand interactions. As no interligand interactions can arise for the first ligand addition the condition of symmetry (22) reduces for this case to

$$\delta G_{xx}(4) = 2\delta G_{xx}(3) \qquad \text{(23)}$$

where $\delta G_{xx}(4)$ and $\delta G_{xx}(3)$ represents the total value of the free energy couplings in the molecules with 4 and 3 ligands, respectively. In the scheme of Figure 3 the binding sites of a tetramer are numbered 1 to 4. As shown in the figure there are six possible free energy coupling between ligands bound at the four sites. In the three different models, A, B, and C the intrinsic free energies of binding are given in the same RT units. Case A is an evenly linked model, all couplings being identical. In B the sites are unevenly linked, but the couplings are pairwise symmetric, and both A and B obey Eq. (21b). The simplest case in which this equation is not obeyed is one in which the couplings among the sites do not show symmetry, like C. We are thus led to the conclusion that asymmetric ligand binding requires a corresponding asymmetry in the interactions among the bound ligands and it seems reasonable to expect that this

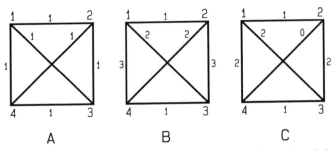

Figure 3. Models of coupling among ligand-binding sites. (A) Uniform coupling; (B) symmetric; (C) asymmetric.

entails in turn some structural protein asymmetry. This appears to be an example of "Curie's law of symmetry" formulated by Pierre Curie in 1893: "When effects are asymmetric, the asymmetry should be apparent in their causes."

Biological Implications of Binding Asymmetry

The titration curves for the case of proteins that bind the ligand at a single site, or at two sites are always symmetric, irrespective in the latter case of the existence of free energy couplings. The simplest protein that can show asymmetry of titration curve is one that binds three molecules of ligand. From Eqs. (21b) we shall define the asymmetry γ as

$$\gamma_3 = \ln\left(\frac{K_2^2}{K_3 K_1}\right); \qquad \gamma_4 = \ln\left(\frac{K_2 K_3}{K_1 K_4}\right) \tag{24}$$

for three-binders and four-binders, respectively. Evidently the asymmetry can be positive or negative according to whether $K_2^2 > K_3 K_1$ ($K_2 K_3 > K_1 K_4$) or $K_2^2 < K_3 K_1$ ($K_2 K_3 < K_1 K_4$) while $\gamma = 0$ corresponds to the symmetry condition. We note that simple binding cannot lead to asymmetry since in that case one uniformly has $\gamma_3 = \gamma_4 = 0$. Also in the three-binder cooperativity requires $K_1/K_2 < K_2/K_3$ and therefore in this case cooperativity cannot coexist with positive asymmetry. The four-binder is the simplest case in which cooperativity can be accompanied by either positive or negative asymmetry. Asymmetry and ligand distributions are closely related: Figure 4 shows the distribution of ligands for a four-binder with cooperativity characterized by the same Hill coefficient, and asymmetry with values -1.52 and $+1.52$, respectively. In the negative asymmetry case, f_1 exceeds the value corresponding to the random distribution at $\langle n \rangle = 1$, while at $\langle n \rangle = 3$, f_3 is smaller than the corresponding random distribution fraction. For positive asymmetry the opposite happens; there is defect of f_1 at $\langle n \rangle = 1$ and excess of f_3 at $\langle n \rangle = 3$ with respect to the corresponding fractions of the random distribution. As a result, in the case of positive asymmetry the cooperativity is most marked for the binding of the third and fourth ligand while in the case of negative asymmetry the cooperativity is most marked for the binding of the first two ligands. When compared with the similar cooperative case with $\gamma_4 = 0$ the titration curve for negative asymmetry leans toward $\langle n \rangle = 0$ while for positive asymmetry it leans against $\langle n \rangle = 4$. This difference is best appraised by a plot of the Hill coefficient against the number bound for the two cases, and this is shown in Figure 4. This plot permits us to gauge the possible biological importance of the asymmetry of the titration curve. It may be exemplified

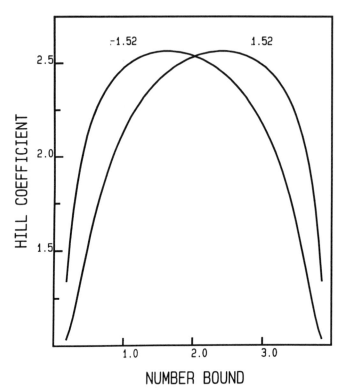

Figure 4. Plots of the Hill coefficient against number bound for a protein with four binding sites and opposite asymmetries, $\gamma = 1.52$ and $\gamma = -1.52$, respectively. $K_1 = 43$, $K_4 = 1$ in both cases. For positive asymmetry $K_2 = 33.89$, $K_3 = 5.82$. For negative asymmetry $K_2 = 7.38$, $K_3 = 1.27$. Notice the increased cooperativity of binding at $N < 2$ if $\gamma < 0$ and at $N > 2$ for $\gamma > 0$.

by hemoglobin, a four-binder of oxygen with high cooperativity and pronounced positive asymmetry (see Chapter VII). Physiologically hemoglobin acts as an oxygen carrier with saturation that drops to some 60% in the tissues and reaches 99% in the lungs. Starting with this latter value the positive asymmetry permits greater unloading of oxygen to the tissues for a given drop in oxygen tension, as compared with the case of no asymmetry and even more so with the case of negative asymmetry. On the other hand in a four-binder, or three-binder, the negative asymmetry would enhance the cooperative effects at low ligand saturation. Therefore the efficiency of an oxygen scavenger would be improved by cooperative binding of oxygen with negative asymmetry.

References

Because of the paucity of appropriate methods of measurement there are few determinations of the distribution of ligands among the binding protein molecules. The following short list probably covers most of the published observations:

1. Viggiano, G., Ho, N.T., and Ho, C. (1979) *Biochemistry* **18**, 5238–5247.

2. Weber, G., and Daniel, E. (1966) *Biochemistry* **5**, 1900–1907. Anderson, S.R., and Weber, G. (1969) *Biochemistry* **8**, 371–380.

3. Lehrer, S.S., and Leavis, P.C. (1978) *Methods Enzymol.* **49**, 222–249.

4. Weber G. (1982) *Nature (London)* **300**, 603–607. Allen, D.W., Guthe, K.F., and Wyman, J. (1950) *J. Biol. Chem.* **187**, 393–410.

VII

Binding by Multimer Proteins

Intracellular proteins are often found to consist of several peptide chains, identical or somewhat different in amino acid sequence, that associate into multimers of fixed composition. Dimers, trimers, tetramers, hexamers and dodecamers have been isolated and characterized. Their prevalence and the belief that the more complex molecular functions of proteins are, and perhaps can only be, performed by them have excited the interest of many biochemists. The progress of X-ray structural analysis of protein crystals has had a pronounced influence in this area by showing that in the particular case of hemoglobin, a tetramer molecule, the oxygen-liganded form has distinct differences in the folding of the peptide chains and their disposition relative to each other when compared with deoxyhemoglobin, the unliganded form [1]. The fact that hemoglobin shows cooperative oxygen binding, as well as conspicuous free energy coupling of the bound oxygen with other ligands like protons and phosphoric esters, has given rise to much speculation as regards the relation of ligand binding and structural changes in oligomeric proteins. This relation between structure and function is still somewhat obscure, as we are presently unable to perform convincing detailed calculations of the magnitude of the free energy couplings starting from the three-dimensional structure of the protein. Leaving aside any implications of the structural data other than the number and kinds of the peptide chains that associate to form the oligomer, a study of the experimental properties of ligand binding and subunit association of proteins in solution (Chapter XV) has made clear that the binding of ligands and the intersubunit interactions are energetically coupled. This free energy coupling can be described in a straightforward manner by the use of concepts already developed in the previous chapters.

Interdependence of Subunit Associations and the Binding of Ligands: Order of the Coupling Effects

We need to distinguish between two different kinds of molecular associations: On the one hand there is the association of the monomers to form the aggregate, and on the other the association of the ligand—a small molecule—with the specific binding sites of the protein. The former type will be called the "macroassociation" and the latter type the "microassociation." Macro- and microassociation processes may be considered on an equal footing: a dimer that binds a mole of ligand can also be considered as a monomer that associates with two ligands: the small molecule and the second monomer. It follows that the thermodynamic description of these processes will not entail anything fundamentally different from the case in which two different ligands, X and Y, are bound to the same protein, a case already discussed in detail. Notice, however, an important difference as regards the first ligand bound by an isolated peptide chain and by a protein aggregate: In the former the standard free energy change on binding of the first ligand bound is, by definition, an unconditional free energy change. In the protein aggregate there can be a free energy coupling between the binding of the first bound ligand and the free energy of association of the ligand-bound subunit with the other subunits in the oligomer. The existence of such couplings is simply deduced from the observations of a difference between two macroscopic quantities: the free energies of ligand binding by the isolated monomers and by the aggregate, or equivalently, by the difference in association free energies of the unliganded and the liganded protein monomers. The existence of a mutual influence between the processes of ligand binding and subunit association is a simple consequence of free energy conservation in the system and does not predicate any particular mechanism for the connection of the two processes; it simply asserts the energetic consequences that follow if such interaction exists, whatever the underlying cause. We can distinguish three types of relations between the processes of ligand binding and subunit association:

1. The two processes are independent. In this case the free energy of association of the monomers is the same whether they are ligand bound or otherwise. Conversely the free energy of ligand binding is identical for the free monomer and the aggregate.

2. Ligand binding induces aggregation: In this case the free energy of ligand binding by the aggregate is larger in absolute value than the free energy of binding by the isolated monomer. Conversely

the free energy of association of the liganded monomers is larger in absolute value than the free energy of association of the unliganded monomers: There is mutual stabilization of oligomer association and ligand binding.

3. Ligand binding induces dissociation. In this case the relations are the opposite: The oligomer binds the ligand more weakly than the monomers and the ligand-bound monomers aggregate into the oligomer less readily than the ligand-free monomers: Ligand binding stabilizes the free monomers.

The varied relations between micro- and macroassociations are best understood by reference to the simplest system, a dimer D made up of identical monomers, A, each of which can bind a single ligand X. ΔG_{11} designates the standard free energy of binding of X to A while ΔG_{21} and ΔG_{22} are, respectively, the standard free energies of microassociation of the first and second moles of X bound to the dimer. $\Delta G(0)$, $\Delta G(1)$, and $\Delta G(2)$ are the free energies of macroassociation that generate the species D, DX, and DX_2 by the respective reactions: A + A, A + AX, AX + AX. The six standard free energies of the possible reactions in the system define completely the relations of macro- and microassociations. These can best be appraised by means of graphs that depict the possible paths for the formation of each possible aggregate of subunits and ligands starting from the isolated components 2A + 2X (Figs. 1 to 5). The horizontal segments in the graphs correspond to the Gibbs free energies of the various molecular species in their standard states and the vertical segments are proportional to the changes in standard free energy in the indicated association reactions. It is easy to see that a chemical species that involves both macro- and microassociations can be generated by carrying out first the macro- and then the microassociations or, alternatively, by inverting the order of these reactions; as the initial and final products are the same the total free energy change must be independent of the order of the operations that generate the latter from the former. These simple considerations of free energy conservation, evident in the figure, give:

$$\Delta G(0) + \Delta G_{21} = \Delta G_{11} + \Delta G(1)$$

$$\Delta G(1) + \Delta G_{22} = \Delta G_{11} + \Delta G(2) \tag{1}$$

The relations (1) reduce the number of independent standard free energies to four. Besides the trivial case of independence of ligand binding and subunit association, characterized by the relations $\Delta G(0) = \Delta G(1) =$

$\Delta G(2)$, there are three qualitatively different cases of interdependence:

1. Binding of the first ligand X changes the free energy of subunit association, while binding of the second does not produce any additional change: $\Delta G(0)\langle\rangle\Delta G(1) = \Delta G(2)$. We shall designate this as a *first-order effect* of ligand binding upon macroassociation.

2. Changes in subunit association take place only when two X-ligands have been bound. In this case, which we designate as a *second-order effect* of ligand binding on subunit association, $\Delta G(0) = \Delta G(1)\langle\rangle\Delta G(2)$.

3. Both ligands affect the subunit association to equal extents, in which case $\Delta G(1) = 1/2[\Delta G(0) + \Delta G(2)]$. This case will be one of *intermediate order*—between the first and the second—of the effects of ligand binding on subunit association.

Any possible relation between subunit interaction and ligand binding by the dimer can be described by mixing effects of either first or second order with those of intermediate order, without adding however any qualitative novelty. In the case of first-order interactions we have two subclasses according to whether $|\Delta G(0)| > |\Delta G(2)|$, in which case ligand binding promotes dissociation or conversely $|\Delta G(0)| < |\Delta G(2)|$, in which case ligand binding promotes association. Inspection of Figures 1 and 2 shows that the former case corresponds to cooperative X-binding and the latter corresponds to antagonistic binding. We can therefore establish as characteristic of first-order effects the correlations:

ligand promoted dissociation = X-binding cooperativity

ligand promoted association = X-binding antagonism

From similar considerations regarding second-order effects we establish precisely the opposite correlations:

ligand promoted association = X-binding cooperativity

ligand promoted dissociation = X-binding antagonism

Therefore by the qualitative determination of the change in aggregation of the subunits on binding, and of the character, cooperative or antagonistic of the X-binding by the aggregate we are able to decide whether the effects of ligand association are of first or second order.

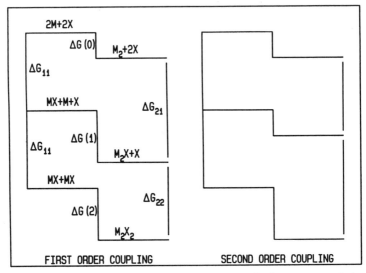

Figure 1. Ligand-induced association can result from antagonistic binding of the ligands (first-order coupling, left) or cooperative binding (second-order coupling, right).

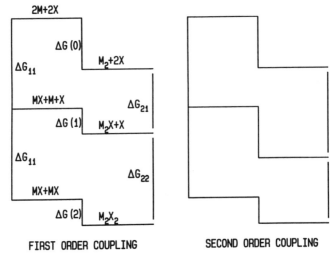

Figure 2. Ligand-induced dissociation can result from cooperative binding with first-order couplings (left) or antagonistic binding with second-order couplings (right).

INTERMEDIATE ORDER COUPLING

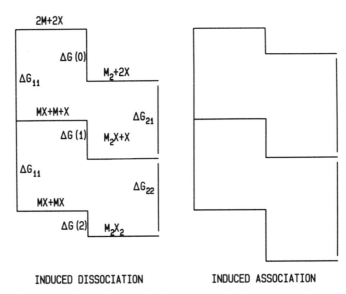

Figure 3. Ligand-induced dissociation (left) and association (right) can result from pure intermediate order couplings when $\Delta G(1) = (1/2)[\Delta G(0) + \Delta G(2)]$. It gives rise to simple binding.

The coexistence of simple binding ($\Delta G_{21} = \Delta G_{22}$) with promoted association or dissociation is characteristic of the effects of intermediate order. The important point follows that if an analysis of the microassociation shows that $\Delta G_{21} = \Delta G_{22}$ it is not possible to conclude that macro- and microassociations are independent (Fig. 3). A number of cases have been reported in which addition of a ligand produces evident dissociating or associating effects on an oligomeric protein without a similar report of differences in the free energies of microassociation. Although many of these cases require a more thorough investigation before being put in this category, theory has certainly a place for them and they may turn out to be common in practice.

We need also to consider the case in which intermediate order effects coexist with those of first or second order. One such case is shown in Figure 4. By comparison with Figure 2 we can easily see that in pure first-order or second-order effects

$$|\Delta G(2) - \Delta G(0)| = |\Delta G_{21} - \Delta G_{22}| \qquad (2)$$

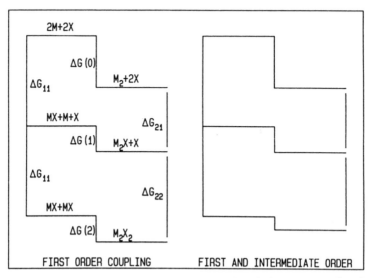

Figure 4. Comparison of ligand-induced dissociation as a result of pure first-order couplings (left) and mixed first- and intermediate-order couplings (right). The cooperativity $(\Delta G_{22} - \Delta G_{21})$ is reduced in the latter case.

whereas in the cases of mixed orders

$$|\Delta G(2) - \Delta G(0)| > |\Delta G_{21} - \Delta G_{22}| \tag{3}$$

This distinction has practical importance: The four quantities appearing in Eqs. (2) and (3) can be, always in principle and in many cases in practice, independently measured. In drawing up the energetic balance shown in Eq. (1) we took the free energy of binding of an isolated subunit as our reference state. The subunit interactions are always the cause, direct or indirect, of the difference between the ligand affinities of aggregate and isolated subunit, and this must be repaid by free energy changes at the subunit boundary. However, a meaningful comparison of ΔG_{11} with ΔG_{21} and ΔG_{22} may not be possible unless measurements of the binding of the ligand by the aggregate and the separated monomers in the same media can be made.

The Monomer as Second Ligand

Considering the monomer as a further ligand that takes the place of Y in the binding of two ligands to a single peptide chain (Chapter III), we

Table 1. Equivalences of Standard Free Energy Changes

Reaction	ΔG	Equivalent to
1. $A + X \rightarrow AX$	$\Delta G(X)$	ΔG_{11}
2. $A + A \rightarrow A_2$	$\Delta G(A)$	$\Delta G(0)$
3. $AX + A \rightarrow A_2X$	$\Delta G(A/X)$	$\Delta G(1)$
4. $A_2 + X \rightarrow A_2X$	$\Delta G(X/A)$	ΔG_{21}
5. $AX + AX \rightarrow A_2X_2$	$\Delta G(AX/X)$	$\Delta G(2)$
6. $A_2X + X \rightarrow A_2X_2$	$\Delta G(X/AX)$	ΔG_{22}

can derive the equivalences of the standard free energy changes shown in Table 1.

The conservation relations

$$\Delta G(A) + \Delta G(X/A) = \Delta G(X) + \Delta G(A/X)$$

$$\Delta G(A/X) + \Delta G(X/AX) = \Delta G(X) + \Delta G(AX/X) \qquad (4)$$

are equivalent to those in Eq. (1). In analogy with the interactions of X and Y bound to a single peptide chain we can define: $\Delta G_{A,X}$, the free energy coupling between the binding of an unliganded monomer and an X-ligand to a monomer, and $dG_{AX,X}$, the coupling between the binding of a liganded monomer and an X ligand. From these definitions and the equivalences of Table 1 we find that

$$\delta G_{A,X} = \Delta G(1) - \Delta G(0) = \Delta G_{21} - \Delta G_{11}$$

$$\delta G_{AX,X} = \Delta G(2) - \Delta G(1) = \Delta G_{22} - \Delta G_{11} \qquad (5)$$

The last two equations describe the reciprocity of the effects between the free energy changes of subunit interaction, at the subunit boundary and of small ligand binding, at the ligand binding site.

Free Energy Couplings in a Heterodimer

The energetic relations of ligand binding and subunit association for a dimer AB made of two different peptide chains are shown in Figure 5. The isolated monomers bind the ligand with standard free energies $\Delta G_{11}(A)$ and $\Delta G_{11}(B)$, which may be appreciably different, thus generating a twofold splitting of the Gibbs level for the liganded monomers: AX + B + X and A + XB + X. Similarly there are two levels for the unliganded dimers XAB and ABX, which determine two different free energies of

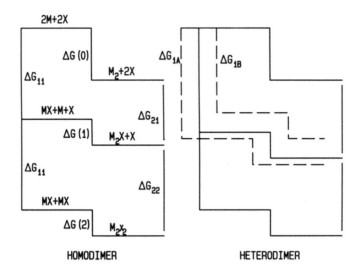

FIRST ORDER COUPLINGS

Figure 5. Comparison of ligand-induced dissociation of a homodimer (left) and heterodimer (right). The cooperativity of ligand binding is reduced as it is by the mixture of first- and intermediate-orders effects (Fig. 4).

association, $\Delta G(1A)$ and $\Delta G(1B)$. The reactions involving one ligand attached to either A or B in the dimer, shown in Figure 5 by discontinuous lines, may be considered separately. We thus have

$$\Delta G_{21}(A) = \Delta G_{11}(A) + \Delta G(1A) - \Delta G(0)$$

$$\Delta G_{21}(B) = \Delta G_{11}(B) + \Delta G(1B) - \Delta G(0) \qquad (6)$$

ΔG_{21} is an average value given by

$$\Delta G_{21} = \Delta G_{21}(A)\, f(A) + \Delta G_{21}(B)\, f(B) \qquad (7)$$

with

$$M = \exp - \left(\frac{\Delta G_{21}(A)}{RT} \right) + \exp\left(\frac{-\Delta G_{21}(B)}{RT} \right)$$

$$f(A) = \frac{\exp(-\Delta G_{21}(A)/RT)}{M}; \qquad f(B) = \frac{\exp(-\Delta G_{21}(B)/RT)}{M} \qquad (8)$$

ΔG_{21} will increase and ΔG_{22} will decrease with respect to the values that they would have if the monomers were identical; as a result cooperative binding effects would be decreased and apparent antagonism increased in comparison with the values that would be obtained in the case of identical dimers. If $\Delta G_{21}(A)$ and $\Delta G_{21}(B)$ differ by $1RT$ or less ΔG_{21} will not substantially differ from the value that is obtained when the monomers are identical, but if they differ by $2RT$ or more ΔG_{21} will be very close to the larger of the two free energies $[\Delta G_{21}(A)$ or $\Delta G_{21}(B)]$, and there will be almost complete precedence of binding by one of the monomers within the dimer. Comparison of Figure 5 with Figure 4 shows that in the relations of macro- and microassociations this case does not formally differ from one of identical monomers with free energy couplings of mixed order. The existence of effects of mixed order indicates that on the formation of the aggregate there appear interactions between ligand binding and subunit interaction that can be considered as belonging to each separate chain: although they affect the subunit association they do not result in any change in the ligand affinity of the binding site of the opposite monomer. The difference between effects of first and second order on one hand and of intermediate order on the other can be schematized as shown in Figure 6. The former correspond to "connections" of a single boundary region with the two binding sites, and the latter to connections of separate boundary regions with each of the two binding sites.

Titration with Ligand of the Dissociating Dimer

Since the protein dimer has a finite free energy of association distinctly different binding phenomena will be observed depending on the protein concentration. At dilutions such that both liganded and unliganded forms are completely dissociated the only relevant free energy of binding of the ligand will be ΔG_{11}. At the other extreme, if the protein concentration is larger than the three macrodissociation constants, $k(0)$, $k(1)$, and $k(2)$, the titration data will correspond to a single macromolecular species, the dimer, which is described by the two Adair constants K_{21} and K_{22}. At protein concentrations intermediate between the dimer dissociation constants of the unliganded and fully liganded dimer the titration of protein with ligand takes a more complex course: The expected species in solution will be A, AX, A_2, A_2X, and A_2X_2 where A designates the monomer. We can set

$$P_1 = 1 + [X]/K_{11}; \qquad [A] + [AX] = [A]P_1$$

$$P_2 = 1 + 2[X]/K_{21} + [X]^2/(K_{21}K_{22}) \tag{9}$$

$$[A_2] + [A_2X] + [A_2X_2] = [A_2]P_2 \tag{10}$$

BINDING SITE-BOUNDARY EFFECTS

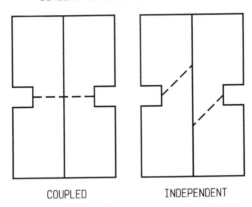

COUPLED   INDEPENDENT

Figure 6. Schematic representation of the binding site boundary connections that permit independence of the effects of binding (right) or their dependence (left) leading to either cooperative or antagonistic effects when the intrinsic free energies of binding at the two sites are the same.

In these equations K_{11} is the dissociation constant of monomer–ligand complex, and K_{21} and K_{22} the Adair constants for the binding of X by the dimer. The dissociation constant of unliganded dimer into monomers equals

$$k(0) = \frac{[A]^2}{[A_2]} \tag{11}$$

If the total concentration of protein is expressed as dimer molarity, D_0

$$D_0 = \frac{P_1[A]}{2} + P_2[A_2] \tag{12}$$

A quadratic equation in $[A]/k(0)$ results if $[A_2]$ is eliminated from Eq. (12) by means of Eq. (11):

$$\left[\frac{[A]}{k(0)}\right]^2 + \frac{P_1}{2P_2}\frac{[A]}{k(0)} - \frac{D_0}{P_2 k(0)} = 0 \tag{13}$$

Its solution is

$$\frac{[A]}{k(0)} = m\left(\sqrt{\left[1 + \frac{D_0}{k(0)m^2 P_2}\right]} - 1\right) \tag{14}$$

where we have written $m = P_1/(4P_2)$. Notice that when the second term under the radical is much smaller than unity, that is when D_0 is sufficiently small,

$$P_1[A] \rightarrow 2D_0 \tag{15}$$

indicating that the monomer forms, liganded and unliganded, account for virtually all the protein. Conversely if the second term under the radical is very large compared to unity,

$$P_2[A_2] \rightarrow D_0 \tag{16}$$

in which case the protein is almost entirely in the dimer forms. Defining the parameter r as the ratio $[A_2]/[A]$, it follows from Eq. (11) that

$$r = \frac{[A]}{k(0)} \tag{17}$$

The average number bound is

$$\langle n \rangle = \frac{2(P_1 - 1) + r(P_2 - 1 + [X]/K_{21})}{P_1 + rP_2} \tag{18}$$

Equations (13) and (17) show how the average number bound may be computed as a function of [X] when the dissociation constants K_{11}, K_{21}, and K_{22} and the parameter $D_0/k(0)$ are specified.

Order of the Free Energy Couplings from Titration Data

Various effects are obtained according to whether the free energy couplings between ligand binding and subunit interaction are of the first, second, or intermediate order. When the couplings are of first or second order without admixture of intermediate order, and the binding is cooperative in character, either $K_{21} > K_{11}$ and $K_{22} = K_{11}$ (first-order coupling) or $K_{21} = K_{11}$ and $K_{22} < K_{11}$ (second-order coupling). Observation of the cooperative character of the microassociation requires high protein concentrations and the Bjerrum plots 2 and 3 of Figure 7 are for $D_0/k(0) =$

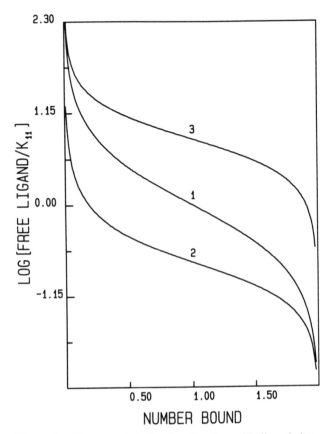

Figure 7. Bjerrum plots for the titration with ligand that binds cooperatively at two sites with a ratio $K_{21}/K_{22} = 50$. Curve 1 is a titration curve at $D_0/k(0) \rightarrow 0$. It corresponds to two monomers with the single dissociation constant $K_{11} = 1$. Curves 2 and 3 are for $D_0/k(0) = 300$. Curve 2: $K_{21} = 50$, $K_{22} = 1$, because of first-order coupling. Curve 3: $K_{21} = 1$, $K_{22} = 0.02$, because of second-order coupling.

300. Plot 1 of this figure is that for a dimer with two independent sites with single dissociation constant K_{11}, equivalent therefore to that of the isolated monomers, observable at "infinite" dilution, when $D_0/k(0) \ll 1$. The plots show that comparison of titration with ligand at low and high concentration of protein is sufficient for a decision as to the existence of cooperativity or antagonism between the bound ligands. A decision as to

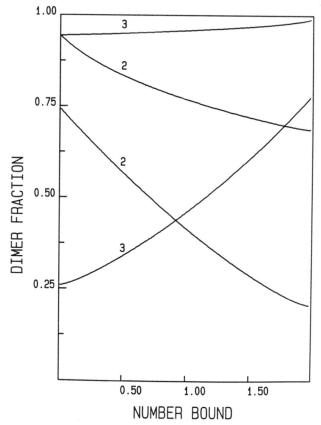

Figure 8. Plots of degree of dissociation against number bound. The two upper curves are for $D_0/k(0)$ = 300. Lower curves: 2, showing dissociation on binding, is for first-order couplings and $D_0/k(0) = 10$; 3, showing association on binding, is for second-order coupling and $D_0/k(0)$ = 0.3.

the order of the couplings in cooperative ligand binding is carried out as shown in Figure 8: ligand addition results in dimer dissociation if coupling is of first order (plots 2) and in monomer association if it is of second order (plots 3). Observation of strong dependence of the monomer association on the ligand association requires observations at a total protein concentration of the same order of the macroassociation constant of the uniliganded species. For the purpose of a more quantitative analysis, fitting of the experimental titration data at various protein concentrations can be used to determine the four free energies appearing in Eqs. (5). Equation (1) may then be used to compute the additional two: $\Delta G(1)$ and $\Delta G(2)$. Independent physical methods may be able to determine $\Delta G(0)$ and $\Delta G(2)$ by studies of the dimer–monomer equilibrium, respectively, in the absence and in the presence of saturating concentrations of ligand. These values can provide a check on those derived from the titration with ligand. With $\Delta G(0)$, $\Delta G(1)$, and $\Delta G(2)$ known we can compute an index

$$I = \frac{\Delta G(2) - \Delta G(1)}{\Delta G(1) - \Delta G(0)} \qquad (19)$$

If $I = 1$ the free energy couplings between subunit interaction and ligand binding are of intermediate order. If $I > 1$ they are of second order or mixed second and intermediate order and if $I < 1$ they are of first order or mixed first and intermediate order.

Extension of the Concept of Order of the Free Energy Couplings

Relatively few oligomeric proteins are dimers; a greater number are tetramers and even larger aggregates. Nevertheless the concept of the order of the free energy couplings may be usefully applied to these larger aggregates, in particular to tetramers like hemoglobin. In an aggregate of N monomers there are $N(N - 1)/2$ possible independent interactions between pairs of subunits, but this number is likely to be diminished by symmetries that result from the presence of identical subunits, and in the larger aggregates from the absence of common boundaries between some of the possible pairs. The effect of twofold symmetry is illustrated by hemoglobin: the six possible interactions among the two α (α_1, α_2) and the two β (β_1, β_2) subunits that form the molecule are of only four types: $\alpha_1\beta_1 = \alpha_2\beta_2$, $\alpha_1\beta_2 = \alpha_2\beta_1$, $\alpha_1\alpha_2$, and $\beta_1\beta_2$. The changes in boundary interaction caused by the binding of ligands to a given pair of subunits can be of the orders that are operative in the dimer case: first, second, intermediate, or a mixture of the latter and one of the former types. In any aggregate larger than a dimer there is the possibility that any change in boundary interaction will require the binding of ligands at more than two subunits, that is free energy couplings between ligand binding and subunit interaction could conceivably be of an order higher than the second. In common with other cases in particle physics, we expect that interactions between pairs will permit us an analysis that will be qualitatively, and often quantitatively correct of the relations between the energetics of ligand binding and subunit interaction. Besides, as described below, the experimental data may allow us to determine the relative importance of pair interactions and those of a higher order. It must be carefully noted that these interactions between pairs refer to the subunits *as they are present in the aggregate*; nothing is implied about those interactions that are operative between the subunits when the aggregate splits into fragments larger than the monomer.

To illustrate the further application of the concepts of the order of the free energy couplings between ligand binding and subunit association and in particular to show that it may be possible to decide whether the couplings are of an order higher than the second, we examine in more detail the case of a tetramer. The chemical reactions involved in the macro- and microassociations that are relevant to the tetramer–dimer equilibria, and their corresponding standard free energy changes are:

1. Ligand additions:

$$D + 2X \rightarrow DX_2: \qquad \Delta G_{21} + \Delta G_{22}$$

$$T + 2X \rightarrow TX_2: \qquad \Delta G_{41} + \Delta G_{42}$$

$$TX_2 + 2X \rightarrow TX_4: \qquad \Delta G_{43} + \Delta G_{44}$$

These three free energies of ligand addition are the analogues of ΔG_{11}, ΔG_{21}, and ΔG_{22}, respectively.

2. Macromolecular associations:

$$D + D \rightarrow T: \qquad\qquad\qquad \Delta G(0);$$

$$DX + DX \leftrightarrow D + DX_2 \rightarrow TX_2: \qquad \Delta G(2);$$

$$DX_2 + DX_2 \rightarrow TX_4: \qquad\qquad \Delta G(4);$$

These three standard free energy changes are the analogues of $\Delta G(0)$, $\Delta G(1)$, and $\Delta G(2)$, respectively.

Although tetramer–dimer and dimer–monomer equilibria are in most respects similar, there is one important difference: The biliganded tetramer may be generated by two different reactions: $DX + DX \rightarrow TX_2$ and $D + DX_2 \rightarrow TX_2$. Therefore, at equilibrium TX_2 will coexist with a mixture of the unliganded, singly liganded, and doubly liganded dimers. The relative proportions of these dimers, and therefore the average free energy of the mixture, is determined by the disproportionation equilibrium

$$D + DX_2 \leftrightarrow DX + DX$$

This has equilibrium constant

$$K_{dp} = \frac{[D][DX_2]}{[DX]^2} = \frac{K_{21}}{K_{22}} \tag{20}$$

and is therefore determined by the Adair dissociation constants of DX_2. The corresponding free energy of disproportionation is

$$\Delta G_{dp} = \Delta G_{22} - \Delta G_{21} \tag{21}$$

and the average free energy of the mixture of liganded and unliganded dimers in equilibrium with TX_2 is

$$\Delta G_{2m} = 0 \, f(D) + \Delta G_{21} \, f(DX) + (\Delta G_{21} + \Delta G_{22}) \, f(DX_2) \tag{22}$$

where $f(D)$, $f(DX)$, and $f(DX_2)$ are the fractions of the total dimer present as unliganded, uniliganded, and biliganded dimer, respectively. Then, with

$$M = 1 + \exp\left(\frac{-\Delta G_{21}}{RT}\right) + \exp\left(\frac{-\Delta G_{21} - \Delta G_{22}}{RT}\right); \qquad f(D) = 1/M$$

$$f(DX) = \exp\left(\frac{-\Delta G_{21}}{RT}\right) \Big/ M; \qquad f(DX_2) = \exp\left(\frac{-\Delta G_{21} - \Delta G_{22}}{RT}\right) \Big/ M$$

They relate to the free energies of ligand addition by conservation relations (Fig. 9) similar to those appearing in Eqs. (1):

$$\Delta G_{2m} + \Delta G(2) = \Delta G(0) + \Delta G_{41} + \Delta G_{42}$$

$$2(\Delta G_{21} + \Delta G_{22}) - \Delta G_{2m} + \Delta G(4) = \Delta G(2) + \Delta G_{43} + \Delta G_{44} \tag{23}$$

The index I of Eq. (19) becomes here

$$I(4) = \frac{\Delta G(4) - \Delta G(2)}{\Delta G(2) - \Delta G(0)} \tag{24}$$

The value of the index indicates the relative importance of the first and the last two ligands bound as regards the changes in subunit association with ligand binding. It is evident that if $I(4)$ has a value distinctly lower than unity free energy couplings of order higher than the second must be excluded, whereas a value much higher than unity would indicate that couplings of third or fourth order are predominant. We will find that these considerations are indispensable in determining the order of free energy couplings between oxygen binding and subunit association in hemoglobin.

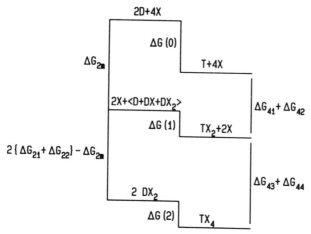

Figure 9. Free energy level scheme depicting the dissociation of a tetramer into dimers, applicable to hemoglobin. D, Dimer; T, tetramer. The significance of ΔG_{2m} is explained in the text.

The dependence of the titration curve of the tetramer that dissociates into dimers on the concentration of protein is essentially similar to that obtained for a dimer that dissociates into monomers.

Physical Origin of Free Energy Couplings

We have already noted that the thermodynamic coupling of the binding of the two ligands, X and monomer, does not predicate any specific mechanism. Nevertheless, if we accept that binding of either ligand is determined exclusively by short-range molecular interactions it follows that binding of X must produce changes at the interacting chemical groups in the subunit boundary and reciprocally that modification of the subunit boundary interactions must reflect itself in the disposition of those groups that make direct contact with the ligand. Whether the changes at the subunit interface will result in simple, cooperative or antagonistic microassociations depends on the intersubunit bonds modified by the microassociation: If the intersubunit bonds formed or broken on microassociation at either monomer are independent of each other no cooperative or antagonistic effects should result, but this effect will follow if certain boundary bonds can be exclusively broken, or formed, by the binding of either one or two ligands (first- and second-order effects mentioned above). True antagonistic binding in which the first molecule of X bound decreases the affinity for the second is not easily distinguished experimentally from binding at two sites with different unconditional free energies of binding. The structural changes required to produce appreciable difference in affinities are indeed minimal and it is not often that the structure of a protein is known in such detail that one can exclude the second possibility. Slade [3] has shown that evident protein heterogeneity can result in apparent antagonistic behavior of NAD binding by glyceraldehyde-phosphate dehydrogenase: Separation by affinity chromatography yielded two fractions of the enzyme, one with clear cooperative coenzyme binding and another with much less catalytic activity and reduced coenzyme affinity. When these two fractions coexist in comparable proportions the titration data show apparent antagonism among bound ligands, and it is quite evident that this arises as a result of averaging over a heterogeneous population and is not the result of true antagonism between bound ligands. It can be reasonably concluded that although antagonistic binding of one ligand at the multiple sites of an oligomeric protein has often been claimed, no conclusive proof of its existence has been given.

Physical Limitations

The reciprocity effects shown in Eqs. (1) are formally equivalent to those between two ligands bound to the same peptide chain and we expect the physical basis of the reciprocity to be the same. When a ligand is bound to a peptide chain its interaction with contacting groups will set up forces that are transmitted and felt at many points of the protein structure. These forces will produce the largest effects where the bonds along the chain are weakest and such a region is provided by the boundary between subunits, as witnessed by the fact that dissociation and hybridization of oligomeric proteins take place more readily than unfolding and denaturation. Although there is in principle no difference bwtween the effects that are set up in monomeric or multimeric proteins it is easy to see why evolution has led to proteins in which the interacting ligands are bound at separate subunits. In this way the influences can be exercised through the effects on a common boundary, the region most likely to be affected by the changes at the binding site. The intervention of the subunit boundaries in the interaction between bound ligands not only presents possibilities that have been developed in evolution but introduces considerable difficulties in the way to a simple structural interpretation of the effects.

Consider in this respect the number and nature of the boundary contacts of the subunits in hemoglobin. According to Perutz et al. [1] and Bolton and Perutz [4], the interchain atomic contacts number about 100 per chain pair. The difference in subunit interaction energy between oxygenated and deoxygenated hemoglobin amounts approximately to 1200 calories [5]. This small change in free energy is to be explained by the changes in the balance of forces distributed over 100 atomic contacts. The drawing up of such a balance requires not only the determination of the atomic coordinates with a precision well beyond that obtained in present day X-ray analysis, but also a very precise knowledge of the relations between displacement of the structure and energy change. It seems that the boundary changes that follow ligand binding present us with a problem where the uncertainties may have significance that goes beyond the technical shortcomings of the present. The classical indetermination of physics arises in cases of large energies and very small particles. Here we are at the other end of the scale, where very small energies are distributed over a very large number of particles. It would, therefore, not be very surprising that the impossibility of measuring all the distances and their energy dependencies with the required precision would limit our knowledge of these phenomena in a more permanent fashion. A discussion of the uncertainties in the measurement of very small distances is given by Brillouin [6]. Although there is no strict upper limit to the precision that may be attained, the "cost" of the information rises prohibitively as the distances fall below atomic dimensions. It may well be that the simultaneous determination of a large number of such distances, with the precision required to draw up a precise energetic balance, will be outside our possibilities or may be our wish. In any case there is little hope that examination of a relevant mutant protein, or a particularly clear experimental result in the measurement of a bulk property will reveal to us the precise "mechanism" of the interaction between ligands bound to a protein. Evidently the probability of a reasonable estimate as to the ligand–protein interactions responsible for a given effect by means of a variational procedure that seeks to minimize the total free energy of the ensemble will rise with the total free energy involved: It is instructive to contrast in this respect the change in subunit interactions in hemoglobin just discussed with the interactions in the complex of chymotrypsinogen with the specific inhibitor [7]. Whereas the former amounts to little more than 1 kcal per subunit the latter involves a free energy change of formation of some 15 kcal [8].

References

1. Perutz, M.F., Muirhead, H., Cox, J.M., and Goaman, L.C.G. (1968) *Nature* (*London*) **219**, 131–139. Baldwin, J.M. (1975) *Prog. Biophys. Mol. Biol.* **29**, 225–320 gives a summary of structure and function of hemoglobin with comprehensive literature on the subject.

2. Noble, R.W. (1965) *J. Mol. Biol.* **39**, 479–491. Weber, G. (1972) *Biochemistry* **11**, 864–878. Weber, G. (1975) *Adv. Protein Chem.* **29**, 1–83. Ackers, G.K. Shea, M.A., and Smith, F.R. (1983) *J. Mol. Biol.* **170**, 223–242.

3. Slade Gennis, L. (1976) *Proc. Natl. Acad. Sci. U.S.A.* **73**, 3928–3932.

4. Bolton, W., and Perutz, M.F. (1970) *Nature (London)* **228**, 551–552.

5. Saroff, H.A., and Minton, A.P. (1972) *Science* **175**, 1253–1255.

6. Brillouin, L. (1946) *Science and information theory*, 2nd ed., Chapter XV. Academic Press, New York.

7. Blow, D.M., Wright, C.S., Chukka, D., Ruhlmann, A., Steigeman, W., and Huber, R. (1972) *J. Mol. Biol.* **69**, 137–144.

8. Green, N.M. (1957) *Biochem. J.* **66**, 407–415

VIII

Hemoglobin

Historical

The only protein-binding system that has been studied in as much detail as the available techniques permit is hemoglobin, the oxygen-carrying blood pigment present in high concentration in the red cells of many animal species. Because of its importance in human physiology and pathology, normal human hemoglobin, hemoglobin A, has been subjected to the most extensive studies. Hemoglobin A is a tetramer formed by two kinds of subunits, α and β, related by their molecular properties, amino acid sequence, and ability to bind oxygen to the single-chain heme protein of muscle, myoglobin. Most of the complexities arising from interactions between subunit association and ligand binding and from interactions among the bound ligands were demonstrated in hemoglobin for the first time. In 1920 Henderson [1] explained the initial observation of Bohr *et al.* (1904) [2] that carbon dioxide repressed the oxygenation of hemoglobin by the fact that oxyhemoglobin is a stronger acid than the deoxygenated protein. We describe this property in the language of ligand interactions by stating that the binding of protons and oxygen to hemoglobin is antagonistically coupled. The binding of oxygen to hemoglobin furnished the first example of cooperative ligand binding and an early formulation of such cooperative binding equilibrium led Hill in 1910 [3] to propose a plot of the data that yields what we now call the Hill coefficient. In 1925 Adair [4] proposed a formulation of oxygen binding in terms of successive oxygen additions, each with a defined equilibrium constant. A generalization of the Adair procedure forms the indispensable thermodynamic basis for the description of multiple binding by proteins. Likewise, the relations between subunit aggregation and ligand binding were first perceived in hemoglobin [5], although the realization of the importance of these relations for the oxygenation of the intact tetramer is much more recent.

It was known from 1925 that blood red cells contain enough 2,3-diphosphoglycerate to form a 1:1 complex with the hemoglobin tetramers [6], but it was only in the 1960s that the observations of Benesch and Benesch, and of Chanutin and Curnish [7] showed the importance of these compounds in controlling the oxygen affinity of hemoglobin through the antagonistic binding of this phosphoric ester and oxygen to the protein. Perutz and his collaborators [8] determined the structure of both oxygenated and deoxygenated hemoglobins in the crystalline form, and this work has had the greatest impact on the development of the subject: On one hand by giving an enormous amount of microscopic information it permitted visualization of all manner of possible relations between the hemoglobin properties and structure, but on the other it discouraged the purely energetic approaches by making it possible to offer for every property a plausible, but almost always weak, inferential explanation based on one or another structural detail [9]. As the X-ray analysis of crystals recognized originally only the wholly deoxygenated and the fully liganded protein states, and these two corresponded to obviously different quaternary structures of hemoglobin, a two-state description of the system was postulated [10]. These two states were assumed to be the only relevant conformations of hemoglobin and to contain all the information necessary to describe the oxygen ligation and the relation of this to other molecular functions. The two-state formulation was readily accepted by many biochemists. On grounds of the better known dynamics of small molecules one can question the oversimplification involved in assuming that a molecule with several thousand potential degrees of freedom can exist in only two relevant states, and as we shall see in the following this latter proposition is flatly contradicted by the analysis of the energetics of oxygen binding. Only after many observations had shown the existence of a very complex dynamics in proteins has the belief in a simple correspondence of structural and energetic states begun to give place to a more flexible approach.

Scope of an Energetic Explanation

A complete description of the energetics of hemoglobin, or any other oligomeric protein of similar size and complexity, is well-nigh impossible. It would involve not only the determination of the energetic couplings of any number of ligands with each other and with the subunit interactions but also the variations in these quantities with pH, temperature, and pressure. It seems sensible therefore to restrict attention to those hemoglobin equilibria that have direct physiological interest, or that because of their simplicity are most likely to reveal unequivocally some important molecular property. Since the properties of hemoglobin are dependent on the ionic composition of the medium it would be important to define one standard buffer medium in which the binding of ions by the protein would be minimal. We cannot define this ideal noninteracting solvent but only eliminate those components that experience has shown to be protein interactive, Tris-HCl buffers of low and medium ionic strength (< 0.2 *M*) are commonly assumed to approach the condition of minimal interaction with hemoglobin and other proteins. The phosphoric esters present in the red cells are strong modifiers of hemoglobin functions and as a point of departure in the determination of their effects one should study the oxygen–hemoglobin equilibria in their absence. After elimination of the phosphoric esters by various methods the hemoglobin solution is referred to as "stripped hemoglobin" [11]. Its study permits us to define the multiple binding of oxygen by hemoglobin in the absence of other significant interactions. Comparison of the equilibria in stripped hemoglobin with those observed in the presence of various "effectors" should then lead to an understanding of the changes brought about by the physiological regulators, the phosphoric esters. In discussing the energetic schemes that couple the macro- and micromolecular interactions we shall attempt to satisfy the experimental data that describe the cooperativity of oxygen binding, the asymmetry of the titration curve, and the dissociation of hemoglobin into identical dimers promoted by oxygen binding. An explanation that satisfies all these aspects of the experimental behavior of hemoglobin leaves very little place for differing interpretations. However, it leaves entirely open the question as to the possible microscopic events that relate the energetics to the structure of the molecule of hemoglobin in various states. Therefore, it seems a logical procedure to attempt a model-free resolution of the energetic aspects before proceeding to an identification of the structural details that are responsible for them. The possibility of, and the necessity for, a model-free consistent description of the energetics in hemoglobin and other cases of complex ligand–protein interactions has not been duly appreciated, with the result that the literature on hemoglobin is full of unproven, and often fanciful correlations of structural aspects and energetic parameters.

Oxygenation Equilibrium of Hemoglobin

This property can be grossly defined by means of the oxygen concentration and the Hill coefficient at midsaturation and many important observations on the properties of hemoglobin in solution do not go further than that. A more complete characterization requires the determination of the four Adair constants for the binding of oxygen. These constants are extracted through computer fitting of the titration curve to the four dissociation constants entering in the Adir equation for binding of a ligand at four sites [12]. Determination of the Adair constants that characterize the binding of oxygen *by the tetramer* necessitates the use of hemoglobin solutions sufficiently concentrated so that dissociation into dimers on oxygen ligation may be disregarded. The standard deviations of the Adair constants thus determined are on the order of $\pm 40\%$, and as would be expected the dissociation constants for the first and the last oxygen molecules bound, K_{41} and K_{44}, respectively, are determined with smaller error than the intermediate dissociation constants, K_{42} and K_{43}. Notice also that an error of 40% implies an indetermination of binding free energy of only ~ 0.5 kcal/mol, or some 7% of the absolute value of the standard free energy of association of oxygen to the protein. In a reversible oxygenation reaction the dissociation constant equals the free oxygen concentration at which midsaturation is reached, $[O_2]_{1/2}$. In writings of physiology this parameter is most often given as $p_{1/2}$ the partial oxygen pressure in mm Hg in equilibrium with hemoglobin at half saturation. $p_{1/2}$ is converted into $[O_2]_{1/2}$ by remembering that at room temperature 1 mm Hg of oxygen pressure results in an equilibrium concentration of oxygen in water or dilute buffer solution of 1.31×10^{-6} M. The Adair constants quoted in the following tables correspond to the oxygen pressures, in mm Hg, at which half of the species $Hb[O_2]_{i-1}$ is converted into $Hb[O_2]_i$. This choice of units makes it possible to perceive directly the physiological significance of the successive steps of oxygen binding. Table 1 shows the constants for stripped hemoglobin obtained in Tris-HCl buffers with and without addition of 0.1 M NaCl, and Table 2 shows some of the literature values of the Adair constants for stripped hemoglobin with various effectors, and of hemoglobin as obtained from the lysis of the erythrocites.

Table 1. Adair Dissociation Constants of Stripped Hemoglobin A[a,b]

	pH	K_1	K_2	K_3	K_4	γ
1.	7	12.6	3.9	1.3	0.23	0.58
2.	7	11	7.6	1.2	0.31	0.6
3.	7	15	4.6	1.7	0.31	0.3
4.	7	49	18	15	0.18	1.1
5.	7	69	96	1.4	0.31	2.0

[a]$K_1 \ldots K_4$ are given as oxygen pressures, in mm Hg. Asymmetry, $\gamma = \ln[K_2 K_3 / (K_1 K_4)]$.

[b]From ref. [12]. 1: Tyuma *et al.* (1971); 2–4: Tyuma *et al.* (1973); 5: Mills *et al.* (1976). 1–3: Tris buffer; 4,5: Tris buffer with 0.1 M NaCl. Note the increase in asymmetry in 0.1 M NaCl, perhaps related to chloride ion binding.

Table 2. Adair Constants of Hemoglobin in Solution under Various Conditions[a]

	pH	K_1	K_2	K_3	K_4	γ
1.	7	125	27	50	0.23	3.29
2.	7	79	28	17	0.23	3.26
3.	7	227	100	400	1.8	4.59
4.	9.1	42	14	2.0	0.27	2.04
5.	7	81	53	3.0	0.78	0.94
6.	7	70	20	12	0.43	0.91
7.	7	88	17	25	0.16	3.4
8.	7	266	70	11^b	0.10	3.3

[a]From ref. [12]. 1–3: Tyuma *et al.* (1973). Human hemoglobin with 1:2 mM 2,3-DPG; 2:2 mM ATP; 3:1.7 mM IHP. 4: Roughton *et al.* (1955). Sheep hemoglobin in borate buffer. 5,6: Lyster and Roughton (1965). Human, horse hemoglobins in 0.6 M phosphate. 7: Knowles and Gibson (1976). Human hemoglobin in 0.1 M phosphate. 8: Di Cera *et al.* (1987), 2 mM IHP.

[b]The authors state that best fitting of the curves is obtained by assuming $1/K_3 \to 0$. The value of the table is calculated from the error given by Di Cera *et al.* for the product $2/(3K_1K_2K_3)$. However, K_3 thus calculated is found to correspond to higher oxygen affinity than some other values of K_3 in this table.

Relation of Oxygenation and Subunit Association

As already discussed in the previous chapter cooperative binding can take place in conjunction with either decreased or increased subunit interaction: The association of the α and β chains into a tetramer can provide for binding constraints that are progressively removed as more ligands are bound by the molecule. In this case we expect the macromolecular association to be weakened by oxygen binding. In opposition the liganded state may provide for a stronger macroassociation: As more ligands are bound the intersubunit bonds are strengthened and this free energy contribution confers increasing affinity of binding to the successive molecules of oxygen that are liganded. The difference in free energy of subunit association between unliganded and oxygen-saturated hemoglobin can be used to distinguish between these two possibilities and in the case of hemoglobin the experimental data are unequivocal [13]: Deoxygenated hemoglobin dissociates into dimers at a concentration four orders of magnitude smaller than the fully oxygenated molecule. Additionally, if binding constraints are present and are progressively removed on oxygen binding we expect that the standard free energy of association of the fourth oxygen bound will be equal to, or somewhat smaller than, the free energy of binding of oxygen to an isolated α or β chain. It would be significantly greater than the latter if the cooperativity originated in

subunit association stabilized by oxygen binding. As shown by Olson *et al.*
[14] the former is the case so that these two crucial criteria coincide in the
conclusion that the origin of the cooperativity is to be found in the
existence of binding constraints that progressively disappear as the number of bound oxygen molecules increases. The excess free energy of
macroassociation of unliganded hemoglobin operates as the binding constraint and part of the free energy of association of the oxygen molecule is
employed in overcoming it. We thus have in the molecular domain the
analogue of D'Alembert's principle in the mechanics of macroscopic
bodies. According to this principle part of the applied forces are inopera-
tive, "lost," at the point of application if they participate in a system of
balanced constraints [15]. These "lost forces" have their counterpart in the
diminished free energy of binding of the oxygen molecules first bound in
comparison with those subsequently bound.

Order of the Free Energy Couplings in Hemoglobin

The next step in the analysis of the energetics of oxygen binding must
necessarily be the determination of the fractional loss of binding constraints as successive oxygen molecules are liganded. With this object we
shall extend to the tetramer the criteria that we developed in a previous
chapter to distinguish the order of the couplings between micro- and
macroassociations in a dimer. First, we evaluate the contribution of free
energy couplings of order higher than the second relative to those of first
or second order. As described in the last chapter $I(4)$ [Eq. (**VII.24**)]
provides a quantitative comparison between the effects of the first two and
the last two ligands bound as regards the removal of the binding constraints. If $I(4) \ll 1$ all the binding constraints have been practically
removed after two ligands are bound and therefore there cannot be free
energy couplings between micro- and macroassociation of an order higher
than the second; there is therefore a negligible contribution of reactions
involving other than pairs of subunits. On the other hand, a value of $I(4)$
significantly higher than unity would indicate that concerted reactions
involving three or more subunits are predominant. We shall confine
ourselves to the determination of $I(4)$ in stripped hemoglobin as there are
at present no data on the dimer dissociation of the complexes of hemoglobin with phosphoric esters. From very complete titration data of stripped
hemoglobin with oxygen at various protein concentrations, Mills *et al.* [12]
determined both the four Adair constants for the tetramer ($K_{41} \ldots K_{44}$)
and the two for the hemoglobin dimers (K_{21}, K_{22}). Their observations, as
well as previous work of Gibson and collaborators [16], show that K_{21} and
K_{22} are virtually equal and differ little from the dissociation constants of

the oxygen adducts of the isolated α or β chains. This circumstance removes any ambiguity as to the chemical potential to be assigned to the mixture of the species D, DX, and DX_2 in equilibrium with the biliganded tetramer; as $K_{21} = K_{22}$ the association reactions $D + DX_2 \rightarrow TX_2$ AND $DX + DX \rightarrow TX_2$ have identical standard free energy changes. Ip and Ackers [13] determined the dissociation constants: $k(0)$, of the unliganded and $k(4)$, of the fully liganded tetramer employing a method that is wholly independent of the oxygen titration experiments. This relies on the very large difference in affinity of haptoglobin for the dimer and tetramer forms of hemoglobin [17]. The values for $k(0)$ and $k(4)$ thus determined are in good agreement with previous determinations by other methods [13]. The experimental free energies of ligand addition and dimer association are gathered in Table 3. The expected partition of the dissociating effects of the binding of the first two ligands in comparison with the effects owing to binding the last two is shown schematically in Figure VII.9.

Equation **(VII.23)** permits one to determine $\Delta G(2)$ employing values from Table 3. Introducing the computed $\Delta G(2)$ a value for $\Delta G(4)$ is obtained that may be compared with the experimentally determined $\Delta G(4)$ given in the same table. In this way we obtain, from the values in the first column of Table 3: $\Delta G(2) = -8.2$ and $\Delta G(4) = -7.9$. the value of $\Delta G(4)$ calculated as indicated and the experimental value are in excellent agreement. From $\Delta G(0)$, $\Delta G(2)$, and $\Delta G(4)$ thus derived, and

Table 3. Dimer Association and Ligand Association Free Energies of Stripped Hemoglobin from Various Sources[a]

	0.1 M NaCl		No NaCl	
	M.A.	T2	T2	T1
ΔG_{21}	-8.4			
ΔG_{22}	-8.4			
ΔG_{41}	-5.5	-5.7	-6.2	-6.4
ΔG_{42}	-5.3	-6.3	-6.8	-7.1
ΔG_{43}	-7.8	-6.4	-7.9	-7.7
ΔG_{44}	-8.7	-9.0	-8.7	-8.7
$\Delta G(0)$	-14.2[b]			
$\Delta G(4)$	-8.0[b]			
$I(4)$	0.05	0.29	0.12	0.06
γ	$+2$	$+1.1$	$+0.6$	$+0.3$

[a]Sources given in ref. [12]: MA, Mills and Ackers; T1, Tyuma *et al.* (1971); T2, Tyuma *et al.* (1973).

[b]Ip and Ackers (1976) [13].

Eq. (9) we get $I(4) = 0.05$. Additional calculations of $I(4)$, making use of the values of $K_{41} \ldots K_{44}$ of Tyuma *et al.* (Table 1) and K_{21}, K_{22}, and $k(0)$ of Table 3, are also given in this latter table. The average $I(4)$, 0.13 ± 0.11, indicates no significant participation of effects of order higher than the second in the free energy coupling of ligand binding and dissociation into dimers. Free energy couplings of third or fourth order are therefore excluded for hemoglobin A, limiting the choice to couplings of either first or second order. We note that the exclusion of a concerted high-order reaction for the case in which the subunit interactions act as binding constraints could have been anticipated, for it would imply that these constraints are removed only at the latter stages of oxygen binding, making it thus impossible for these latter stages to have free energies of binding that exceed those of the early stages, or that are equivalent to the free energy of binding by a free subunit (analogue of D'Alembert's principle). Couplings between macro- and microassociations of third or fourth order would be thus incompatible with cooperativity of ligand binding that results from removal of binding constraints; they could happen only when microassociation promotes subunit association.

Having thus excluded the participation of couplings of order higher than the second, we show below that examination of the data gathered in Table 3 is sufficient to decide that only first-order effects are compatible with the experimental data. To this purpose we consider now in detail the energetics of these two coupling orders: If $G_0 \ldots G_4$ designate the free energy levels corresponding to $0 \ldots 4$ ligands bound, we recall that the relation between dissociation constants and Gibbs levels is

$$\ln K_{4J} = (1/RT)(G_J - G_{J-1}), \qquad 1 \le J \le 4 \tag{1}$$

Table 4. Changes in Constraints Following Ligation

Subunits liganded	Remaining constraints	
	First order	Second order
1. (α)	$S + \{\beta\beta\}$	T
1. (β)	$S + \{\alpha\alpha\}$	T
2. ($\alpha\alpha$)	$\{\beta\beta\}$	$T - \{\alpha\alpha\}$
2. ($\alpha_i\beta_i$)	$\{\alpha_i\beta_j\}$	$T - \{\alpha_i\beta_i\}$
2. ($\alpha_j\beta_j$)	$\{\alpha_i\beta_i\}$	$T - \{\alpha_i\beta_j\}$
2. ($\beta\beta$)	$\{\alpha\alpha\}$	$T - \{\beta\beta\}$
3. ($\alpha\alpha\beta$)	0	$S + \{\beta\beta\}$
3. ($\alpha\beta\beta$)	0	$S + \{\alpha\alpha\}$
4. ($\alpha\alpha\beta\beta$)	0	0

If the free energies are expressed in RT units, as we shall do in the following, RT disappears from the equations. In the hemoglobin tetramer there are six possible subunit interactions: $\alpha_1\beta_1$, $\alpha_1\beta_2$, $\alpha_2\beta_1$, $\alpha_2\beta_2$, $\alpha_1\alpha_2$, $\beta_1\beta_2$. The first four are symmetric interactions in that $\alpha_1\beta_1 = \alpha_2\beta_2$ and $\alpha_1\beta_2 = \alpha_2\beta_1$. We designate below these two kinds of interactions as $\alpha_i\beta_i$ and $\alpha_i\beta_j$, respectively. The remaining subunit interactions are the diagonal ones that we call simply $\alpha\alpha$ and $\beta\beta$. To lighten the notation we shall write $\{\alpha_i\beta_i\}, \{\alpha_i\beta_j\}, \{\alpha\alpha\}, \{\beta\beta\}$ for the differences between unliganded and fully liganded free energies of association residing at the corresponding intersubunit boundaries. The symmetric constraints amount to $S = \{\alpha_i\beta_i\} + \{\alpha_i\beta_j\}$ and the total constraints in the unliganded molecule, which determine the Gibbs level in the absence of ligands, are $T \equiv G_0 = 2S + \{\alpha\alpha\} + \{\beta\beta\}$. We also write $\{\alpha\}$ and $\{\beta\}$ for the free energies of oxygen binding to an isolated α or β subunit, respectively. Table 4 gives the constraints that *remain* at each possible state of ligation when the free energy couplings between ligand binding and subunit association are of either first or second order.

The progressive disappearance of the constraints in the two cases is also illustrated in Figures 1 and 2. The decrease in free energy of ligand binding, with respect to that of the free α or β chain, is determined by the

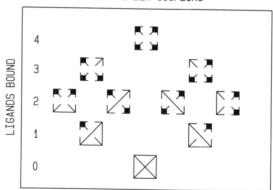

FIRST ORDER COUPLING

Figure 1. Disappearance of binding constraints with number bound for a tetramer in which the free energy couplings are of first order. Solid lines indicate remaining binding constraints; broken lines denote constraints that have disappeared on ligation. Shaded areas indicate liganded sites.

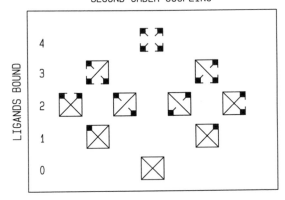

Figure 2. Similar to Figure 1, for free energy couplings of second order.

constraints that disappear when ligation occurs. According to Table 4 binding of the fourth ligand is unconstrained, and therefore equivalent to that of the free chain, in first-order coupling while for second-order coupling binding of the fourth ligand is opposed by three remaining binding constraints. The opposite will happen for the first ligand bound: This stage of ligation will be most constrained, and therefore weakest for first-order coupling and least constrained, and therefore strongest for second-order coupling. Evidently with first-order constraints binding will get stronger as successive ligands are bound while for second-order constraints binding will become weaker at successive stages of ligation. Notice that if the binding of the ligands increased the subunit association precisely the opposite effects would occur: binding of the first ligand will be weakest and binding of the fourth ligand will be strongest for second-order coupling while for first-order couplings binding of the first ligand will be strongest and binding of the fourth will be weakest. Not unexpectedly we find for the tetramer that dissociates into dimers on ligation exactly the same correlations between the order of the couplings and cooperativity or antagonism of ligand binding that are obtained for the dimer that dissociates into monomers on ligation. Generalizing this result we conclude that cooperativity of ligand binding cannot coexist with ligand-induced dissociation if the free energy couplings are of second order, but they necessarily go together if the couplings are of first order. As we have already shown the vanishing importance of couplings of orders higher than the second it follows that the couplings between the free energies of ligand binding and subunit interaction in hemoglobin are of first order.

Asymmetry of the Oxygen Titration Curve of Hemoglobin

We defined the asymmetry of binding in Chapter VI. As a function of the Gibbs energy levels of the tetramer it takes the form

$$\gamma = G_0 - 2G_1 + 2G_3 - G_4 \tag{2}$$

where $G_0 \ldots G_4$, the Gibbs levels of $Hb \ldots Hb(O_2)_4$, are expressed in RT units. We shall examine the dependence of the asymmetry on the six free energies $\{\alpha_i\beta_i\} \ldots \{\beta\beta\}$ that appear in Table 4, and the intrinsic free energies of binding to the isolated subunits, $\{\alpha\}$ and $\{\beta\}$. For first-order couplings we have

$$G_1 = S + (\{\alpha\alpha\} - \{\beta\})f_\beta + (\{\beta\beta\} - \{\alpha\})f_\alpha \tag{3}$$

where f_α and f_β are the fractions of molecules liganded at the α and β subunits. The opposite signs given to the free energies of ligand binding and subunit interaction indicate that the latter are binding constraints. Evidently

$$f_\alpha = \frac{\exp - (\{\beta\beta\} - \{\alpha\})}{\exp - (\{\beta\beta\} - \{\alpha\}) + \exp - (\{\alpha\alpha\} - \{\beta\})}$$

$$f_\beta = \frac{\exp - (\{\alpha\alpha\} - \{\beta\})}{\exp - (\{\beta\beta\} - \{\alpha\}) + \exp - (\{\alpha\alpha\}\{\beta\})} \tag{4}$$

We introduce the differences $\{\alpha\} - \{\beta\} = dI$ and $\{\alpha\alpha\} - \{\beta\beta\} = dD$, which simplify the last equations to

$$f_\alpha = \frac{\exp(dD + dI)}{1 + \exp(dD + dI)}; \quad f_\beta = \frac{1}{1 + \exp(dD + dI)} \tag{5}$$

Also,

$$G_3 = -\{\alpha\} - \{\beta\} - f_{\alpha\alpha\beta}\{a\} - f_{\alpha\beta\beta}\{\beta\} \tag{6}$$

with

$$f_{\alpha\alpha\beta} = \frac{\exp(dI)}{1 + \exp(dI)}; \quad f_{\alpha\beta\beta} = \frac{1}{1 + \exp(dI)} \tag{7}$$

and

$$G_4 = -2\{\alpha\} - 2\{\beta\} \tag{8}$$

Introducing these values of the Gibbs levels in Eq. **(2)** gives

$$\gamma = \{\alpha\alpha\}(1 - 2f_\beta) + \{\beta\beta\}(1 - 2f_\alpha)$$

$$+ 2\{\alpha\}(f_\alpha - f_{\alpha\alpha\beta}) + 2\{\beta\}(f_\beta - f_{\alpha\beta\beta}) \tag{9}$$

With the identities

$$1 - 2f_\alpha = 2f_\beta - 1; \qquad f_\beta - f_{\alpha\beta\beta} = f_{\alpha\alpha\beta} - f_\alpha$$

Eq. **(9)** becomes

$$\gamma = dD(2f_\alpha - 1) + 2dI(f_\alpha - f_{\alpha\alpha\beta}) \tag{10}$$

Introducing the values of f_α and $f_{\alpha\alpha\beta}$ in **(5)** and **(7)**, respectively, into Eq. **(10)**, we conclude

$$\gamma = \frac{dD[\exp(dI + dD)]}{\exp(dD + dI) + 1} + \frac{2dI \exp(dI)[\exp(dD) - 1]}{[\exp(dD) + 1][\exp(dD + dI) + 1]} \tag{11}$$

The preceding derivation assumes that the free energy couplings are of first order. A similar calculation shows that for second-order couplings the asymmetry, γ', is given by Eq. **(11)** preceded by the minus sign, that is $\gamma' = -\gamma$. According to this last equation $\gamma = \gamma' = 0$ if $dD = 0$ whether $dI <> 0$ or $dI = 0$. *Asymmetric binding requires asymmetric free energy couplings between micro- and macroassociations, whether the α and β chains have equivalent affinities or otherwise*, as demanded by Curie's symmetry principle: When the effects are asymmetric the asymmetry should be present in their causes.

If we set $dI = 0$ in Eq. **(11)** we find

$$\gamma = -\gamma' = \frac{dD \exp(dD - 1)}{\exp(dD + 1)} \tag{12}$$

In this case the asymmetry from first-order couplings is always positive and from second-order couplings is uniformly negative. If both dI and dD are appreciable with respect to RT, Eq. **(11)** shows that it is possible for first-order couplings to generate negative asymmetry and for those of second order to generate positive asymmetry, as demonstrated in the plots

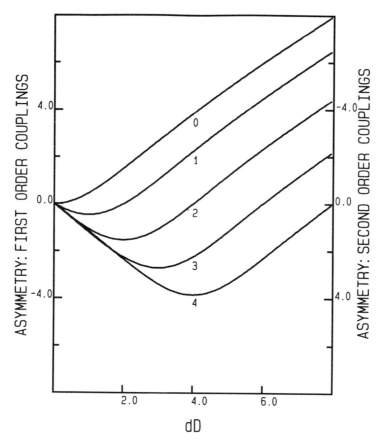

Figure 3. Plots of asymmetry as a function of dD, the difference in the diagonal constraints for first- and second-order couplings. Numbers on the curves indicate dI, the difference in the affinities of the isolated α and β chains in RT units.

of γ against dD of Figure 3. In these cases an inversion of the sign of the asymmetry takes place at $dD + 2dI = 0$. Stripped hemoglobin shows a small positive asymmetry, ~ 0.5, but in the presence of saturating concentrations of various phosphoric effectors the asymmetry can increase to about $+4$. Such large positive asymmetries can be reconciled with second-order couplings only if dI is several times the thermal energy while a value of dI close to RT is consistent only with free energy couplings of the first order. As the isolated α and β chains have nearly equal affinities for oxygen [18] the large positive asymmetries observed in hemoglobin complexed with phosphoric effectors lead to the conclusion that the couplings between macro- and microassociations are of first order. We independently reached the same conclusion by an examination of the relations of cooperativity and subunit aggregation in stripped hemoglobin and this double verification should remove any doubts about its validity.

Derivation of Binding Constraints from the Adair Constants

Whereas it is always possible to derive the Adair constants from a set of free energy couplings, the inverse operation, the determination of a unique set of free energy couplings starting from the four Adair constants, is not in general possible. We can, however, assign constraints that generate, within an acceptable variance, the experimental Adair constants through a suitable computation. Such a computation involves assumptions and steps as follows:

1. Qualitatively and even in a reasonably good quantitative approximation the free energies of binding of the chains $\{\alpha\}$ and $\{\beta\}$ can be taken as equal.

2. If the last oxygen binding step is unconstrained as is the case of stripped hemoglobin in which $\Delta G_{44} = \{\alpha\} = \{\beta\}$, the total value of the binding constraints must equal, in RT units

$$T = \ln K_1 + \ln K_2 + \ln K_3 - 3 \ln K_4 \qquad (13)$$

3. As we demonstrated above the free energy couplings between the macro- and microassociations are of first order. Can the binding constraints in stripped hemoglobin include a contribution from an effect of intermediate order? If each of the Gibbs levels receives an additional identical contribution from changes in subunit interaction occurring at each successive oxygen ligation the total value of the constraints calculated by means of Eq. (13) remains the same. In the presence of effects of intermediate order the total value of the constraints, in stripped hemoglobin, would be significantly smaller than the difference in the free energy of dimer association of unliganded and fully liganded hemoglobin, $\Delta G(0) - \Delta G(4)$. Since these two quantities are equal within experimental errors ($\sim 9.2 \pm 0.7 \, RT$) we conclude that in stripped hemoglobin there are no significant contributions from effects of intermediate order. This result is indeed highly significant: It implies that all the binding constraints residing at the subunit boundaries are shared by the two neighboring subunits; no binding constraint is exclusively connected to one binding site.

In distinction from stripped hemoglobin there are no appropriate experimental data relating tetramer–dimer dissociation and oxygen ligation in the hemoglobin–effector complexes, data that would permit one to derive the order of the free energy couplings in these cases. However, the large positive asymmetries observed in the presence of phosphoric effectors indicates that first-order coupling is operative also in these instances.

4. The difference between the diagonal constraints $dD = \{\beta\beta\} - \{\alpha\alpha\}$ is computed from the experimental asymmetry $\gamma = \ln(K_2K_3/K_4K_1)$ solving by iteration Eq. (12), if it is assumed that $dI = 0$, or Eq. (11) if a specified value for dI is introduced.

5. The total value of the symmetric constraints is

$$S = T - dD - 2\{\alpha\alpha\} \qquad (14)$$

By specification of $\{\alpha\alpha\}$ within the limits 0 and $T - dD$ according to the last equation we can partition the constraints into symmetric and diagonal ones. For each such partition the symmetric constraints $\{\alpha_i\beta_i\}$ and $\{\alpha_i\beta_j\}$ can be given complementary values between the limits 0 and $S/2$.

Although it is not possible from the energetics alone to determine which of the two values of the symmetric constraints should be assigned to $\{\alpha_i\beta_i\}$ or $\{\alpha_i\beta_j\}$ the derived Adair dissociation constants depend—albeit feebly—on their ratio.

The set of binding constraints that best satisfies the experimental data is selected by minimizing the difference between the experimental Adair constants and those derived from the sets of constraints. These assumptions and steps have been used to derive the binding constraints corresponding to the sets of experimental Adair constants displayed in Tables 1 and 2 and the results are shown in Table 5. Particularly in those cases in which the coefficient of variation between calculated and experimental constants is small the experimental constants can be fitted, within their large experimental errors, equally well by different partitions of the symmetric constraints. The uncertainties in the experimental constants deprives of significance any *exact* agreement between their experimental values and those derived from the calculated constraints. However, some qualitative conclusions can be reached when the constraints derived from the various values of the Adair constants found in the literature are compared. Thus, evident differences are observed between stripped hemoglobin and hemoglobin in the presence of the phosphoric effectors, given in Tables 1 and 2: The average symmetric constraints for stripped hemoglobin amount to 4.1 ± 1.1 *RT* and for effector hemoglobin they equal

Table 5. Binding Constraints Derived from the Sets of Adair Constants
of Tables 1 and 2, by the Procedure Described in the Text[a]

	$\{\alpha_i\beta_i\}$	$\{\alpha_i\beta_j\}$	$\{\alpha\alpha\}$	$\{\beta\beta\}$	CV
1.1	1.49	2.23	0	1.1	< 0.01
1.2	1.65	1.65	0	1.52	0.12
1.3	0.96	2.23	0.41	1.48	< 0.01
1.4	2.59	3.11	0	2.26	0.74
2.1	2.29	2.29	1.64	5.63	0.02
2.2	0.5	4.58	0.75	4.22	< 0.01
2.3	2.04	2.04	0.71	5.38	0.30
2.4[b]	2.11	2.11	0.55	2.00	0.01
2.5	1.93	1.93	0.5	1.98	0.32
2.6	0.37	3.30	1.22	3.68	0.02
2.7	0	4.61	1.60	5.20	0.02

[a]The first column contains the table number and observation number within these tables. CV is the average coefficient of variation of the constants K_1, K_2, K_3, derived from the constraints and the corresponding experimental values.

[b]The constraints found for this case (hemoglobin in borate buffer, pH 9.1) are not included in the comparison of stripped hemoglobin and hemoglobin with phosphoric effectors discussed in the text. Stripped hemoglobin has the highest affinity for oxygen, and shows also the smallest asymmetry ($\sim +0.5$).

$4.3 \pm 0.9 \, RT$, showing that these constraints are not significantly changed by addition of the effector. In opposition the diagonal constraints increase from $1.7 \pm 0.5 \, RT$ (stripped hemoglobin) to $4.7 \pm 1.9 \, RT$ (IHP complexes) substantiating the expected cross-linking action of the beta subunits by the phosphoric esters [19].

Table 6 compares the effects of changes in the distribution of constraints on the Adair constants. It illustrates the fact that primarily these constants are determined by the total value of the constraints and by the asymmetry, which is itself uniquely determined by the difference in the diagonal constraints. Changes in the distribution of constraints among the boundaries are relatively unimportant if total value and asymmetry are maintained. On the other hand changes in the constraints that increase their total value or that change the asymmetry have a marked effect on the oxygen titration curve. The greater part of the constraints are in the symmetric interactions, so that total constraints and asymmetry respond to changes in different subunit interactions, thus allowing their independence in physiological control and genetic selection.

It is to be noted that calculations that employ as a starting point the protein structure and the potential functions for atomic interactions, as

Table 6. Effects of Changes in Constraints on the Adair Constants.[a]

	Constraints		K_1	K_2	K_3	K_4	γ
1.	2;2;0;1	(9)	20.3	5.63	1.62	0.28	0.46
2.	2;1;1;2	(9)	20.3	5.70	1.60	0.28	0.46
3.	2;1;0;3	(9)	6.58	8.75	3.30	0.28	2.72
4.	2;1;0;1	(7)	7.47	2.75	1.22	0.28	0.465
2.	5;2.5;0;1	(11)	55.2	9.09	2.74	0.28	0.46

[a]The first four figures are, respectively, $\{\alpha_i\beta_i\}$, $\{\alpha_i\beta_j\}$, $\{\alpha\alpha\}$, and $\{\beta\beta\}$. In parentheses are the total constraints in RT units. The Adair constants are in mm of oxygen pressure. 1: Original distribution of constraints. 2: Shift of $2RT$ from symmetric to asymmetric constraints; total constraints and asymmetry kept constant. 3: Shift of constraints as in 2, but asymmetry increased. 4: Decrease in the absolute value of the symmetric constraints of $2RT$ with unchanged asymmetry. 5: The same: Increase in the symmetric constraints.

presently known, cannot achieve the precision necessary to individualize the various binding constraints, a task that—to a certain extent—we have just shown to be possible by interpretation of the experimental energetics: The subtleties involved in the complex interactions of the subunits within the aggregate exceed by far our present capacity to derive them from microscopic parameters.

Precedence of Binding at the α or β Chains

The Adair constants cannot give us the proportions of liganded α and β subunits as a function of number bound, but the latter information can be obtained from the set of constraints that generates the valid Adair constants. The differences in occupancy of the α and β subunits when one and three subunits are liganded, respectively, are given by Eqs. (5) and (7). A similar derivation can be carried out to yield the relative population of the four levels of G_2, and therefore the α and β occupancies in biliganded hemoglobin. The total α and β occupancies of molecules with 1 to 3 oxygens bound at any value of the number bound is given by the equations

$$n(\alpha) = f(1)f_\alpha + f(2)\left[2f_{\alpha\alpha} + f_{\alpha\beta}\right] + f(3)\left[2f_{\alpha\alpha\beta} + f_{\alpha\beta\beta}\right]$$

$$n(\beta) = \ldots \tag{15}$$

To quantitate the preference of binding I have excluded from Eq. (15) the contribution from the fully liganded molecules, as in this latter population

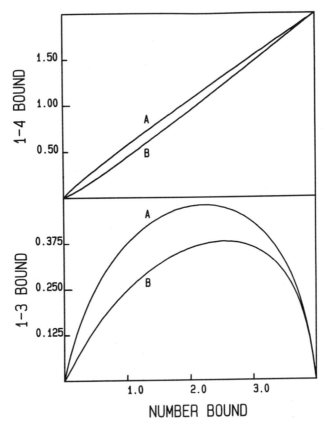

Figure 4. Plots of $n(\alpha)$, $n(\beta)$ of Eq. **(24)** for stripped hemoglobin. If molecules with up to four ligands bound are counted (upper plot) then $n(\alpha) = n(\beta) = 2$ at saturation. If molecules with four ligands are excluded (lower plot) $n(\alpha) = n(\beta) = 0$ when four ligands are bound.

there can be no biased binding, and this is shown in the lower panel of Figure 4. If $n(\alpha)$ and $n(\beta)$ include molecules with four ligands bound the plots take the form shown in the upper panel of the same figure. We note that when molecules with four ligands are included in the computation the precedence of binding becomes much less evident, and that this plot corresponds to what may be expected under the experimental conditions, since then the totality of the molecules bound at each chain is likely to be recorded. It is evident that in stripped hemoglobin there is no appreciable precedence of binding, in agreement with the NMR observations of Ho and co-workers [20].

Physiological Properties and the Energetics of Hemoglobin

Hemoglobin operates as an oxygen carrier between the lungs, or similar respiratory organs, and the tissues. Its saturation varies ordinarily from over 98% in the former to more than 60% in the latter. The changes in the oxygen–hemoglobin affinity in this high saturation region are the ones that should more directly concern us in opposition to the changes that may be observed in the lower saturation range, which are of no direct physiological import. Three parameters are of interest in this respect: (1) The absolute oxygen affinity of hemoglobin. (2) The cooperative character of the last half of the titration curve. (3) The sign and magnitude of the asymmetry. Whereas the physiological worth of the cooperativity of the hemoglobin oxygenation was noticed early, the importance of the asymmetry of the titration curve in the physiological deoxygenation of hemoglobin was recognized by Peller [21] only as recently as 1982. The three parameters just mentioned are not independent of each other and their interplay is best discussed in relation to the effects of the phosphoric esters on the energetics of hemoglobin. The four dissociation constants of stripped hemoglobin are considerably smaller than those determined directly in solutions of hemoglobin obtained by simple lysis of the red cells or those in the intact erythrocytes (Tables 1 and 2). The discovery that anions in general and some organic phosphate esters in particular produce a large decrease in the overall oxygen affinity of hemoglobin opened a new chapter in the understanding of the regulation of oxygen transport in the body. Additionally it permitted visualization of the manner in which the so-called allosteric effectors, molecules structurally unrelated to the substrates, control the binding of the latter substances by enzymes.

A variety of phosphoric esters serve as physiological effectors to regulate the affinity of hemoglobin: 2,3-diphosphoglycerate (DPG) in mammals, ATP in reptiles, inositol hexaphosphate (IHP) in birds. The binding antagonism with oxygen seems to have only a moderate specificity; a variety of polyanions (e.g., the benzene polycarboxylates [22]) produce similar or even greater depressions of the affinity for oxygen. Benesch *et al.* [23] found that on reaction of hemoglobin with 1 mol of formyl pyridoxal phosphate the β chains become covalently cross-linked. The resulting hemoglobin shows high cooperativity (Hill coefficient ~ 2) and permanently decreased oxygen affinity. Other β chain cross-linked derivatives of hemoglobin display similar oxygen-binding properties. The binding of DPG and IHP to the isolated β chains has been directly demonstrated by equilibrium dialysis experiments and the interactions with these chains are apparent in the structure of the complex revealed by the X-ray crystallography of the complexes [19]. The kinetic observations of Wiedermann and Olson [24] show that IHP cross-links two β chains: When added to these chains in molar proportion of 1:2 it catalyzes their association into an all-β tetramer, but it markedly retards it when present in 1:1 proportion. An observation of Wells *et al.* [25] makes quite clear that the decrease in affinity for oxygen of the phosphoric effectors is not universally linked to the presence of cooperativity: Hemoglobin from *Sphenodon*, a very primitive reptile, is tetrameric but devoid of cooperativity, yet its oxygen affinity is decreased by the expected one order of magnitude when 1 mM ATP is added.

The Titration Curve of Hemoglobin in the Presence of Effectors

Addition of milimolar 2,3-diphosphoglycerate to hemoglobin solutions produces an increase in $[O_2]_{1/2}$, or $p_{1/2}$, somewhat smaller than one order of magnitude, an exaltation of the cooperativity shown in an increase in Hill coefficient that may reach up to 3.2, and a notable increase in asymmetry, up to a value of $+3.8$ (Table 2). It is possible to envision the antagonistic couplings of the binding of phosphoric esters and oxygen either as direct effects of the regulator on each binding site independently of the others or indirectly through changes in the binding constraints residing at the subunit boundaries. In the first case we expect the resulting binding constraints to be the sum of those operating in stripped hemoglobin plus the effector-binding site couplings. We discussed in Chapter V the modification of multiple binding by a single ligand of another type, which has identical free energy couplings with each binding site. The observations described above of the specific interaction of the phosphoric effectors with the β subunits show that this simple case is not a good approximation to the situation in hemoglobin A, although it could possibly apply to the hemoglobin from *Sphenodon*. We must therefore assume that if there are independent oxygen–effector couplings, these are different for the α and β subunits.

The Gibbs levels for this case are calculated by adding to the binding constraints of Table 4 the free energy couplings between the phosphoric effector, Y, and the oxygen bound to an α or β subunit, respectively $\delta G_{\alpha - Y}$ or $\delta G_{\beta - Y}$. The binding at each site is opposed by the effector independently of the state—bound or free—of the remaining sites, a situation that is equivalent to an additional coupling of intermediate order between boundary and binding site. On the other hand if the effector

Table 7. Experimental and Computed Adair Constants of Oxygen Binding by the IHP–Hemoglobin Complex[a]

	K_1	K_2	K_3	K_4	γ
Experimental constants	227	100	400	1.8	4.6
Fitting to first-order couplings	223	130	306	1.8	4.6
Fitting to stripped Hb + independent O_2–IHP couplings	252	85	23	4.0	0.6

[a]The total binding constraints, T, are practically the same in all three cases, $\sim 14.5RT$.

induces a new set of binding constraints we will assume that, like in the case of stripped hemoglobin, these are of first order, as demanded by the general association of cooperativity and subunit affinity and by the high positive asymmetry. Employing the two alternative assumptions of effector–subunit interactions independent of the intersubunit constraints, or new, modified constraints we attempt to obtain Adair constants that approach the experimentally observed ones. Table 7 shows a comparison of the values of the Adair constants of the *hemoglobin–effector complex* calculated on these two different suppositions with the constants experimentally determined by Tyuma *et al.* for hemoglobin with saturating amounts (1.7 mM) of IHP. In the calculation of the Adair constants under the assumption of independent effector-site interactions the constraints of stripped hemoglobin employed were $\{\alpha_i\beta_i\} = \{\alpha_i\beta_j\} = 2RT$, $\{\alpha\alpha\} = 0$, $\{\beta\beta\} = 1RT$, and $\{\alpha\} = \{\beta\} = -14RT$, and the independent effector–oxygen interactions were $\delta G_{\alpha-\text{IHP}} = 0.5RT$, $\delta G_{\beta-\text{IHP}} = 2RT$.

Table 7 shows that the assumption of first-order couplings permits fitting of the individual constants and the asymmetry well within the rather large uncertainties of their experimental determination. This is evidently not the case when the boundary constraints of stripped hemoglobin are overlaid by independent O_2–IHP interactions. The calculations by this latter method fail always to reproduce the rather large asymmetry, and values of K_2 and K_3 such that $K_3 > K_2$, a feature observed in many sets of Adair constants in the presence of effectors (Table 2), but systematically absent in stripped hemoglobin (Table 1). This feature, which we now know to result from the very different sets of constraints for the binding of oxygen by the α and β subunits when effector is present, led Antonini [26] originally to propose that the oxygenation of hemoglobin was that of two independent dimers, which he expected to be α–β dimers. His supposition was essentially correct, though the dimers in question are evidently two functional dimers, α–α and β–β rather than two structurally identical α–β dimers.

The Adair constants of Table 7 are those for the binding of oxygen to hemoglobin fully saturated with effector. The dissociation constant of the *effector* increases steadily with the oxygenation of hemoglobin as a result of the reciprocal antagonism of the two kinds of ligands. The progressive dissociation of the effector results in hemoglobin molecules with lower oxygen dissociation constants and the consequent increase in oxygen affinity with saturation have effects identical to the increase in cooperativity of oxygen binding as described in Chapter V. This apparent exaltation of the cooperativity of human hemoglobin in the presence of milimolar DPG, amounting to an increase of about 0.5 in the Hill coefficient above that of stripped hemoglobin, depends on the release of DPG from the complex with hemoglobin at sufficiently high levels of oxygen saturation. As discussed in Chapter V such an effect can be observed only over a limited range of DPG concentration, the range comprised between the dissociation constants of the complexes of DPG with unliganded and fully liganded hemoglobin. When DPG is replaced by a similar concentration of IHP the increase in the Hill coefficient is not observed: The affinity of IHP for oxygenated hemoglobin being too high this effector is not released even at close to oxygen saturation. The same effect should be observed if the concentration of DPG could be raised to become very much larger than its dissociation constant from fully liganded hemoglobin.

From these considerations it follows that at each oxygen saturation value the titration curve is a composite of the curves for stripped hemoglobin and hemoglobin–effector complex. Nevertheless, as shown in Chapter V, this titration curve can be represented by four effective Adair constants that are now dependent on the ratio of concentration of effector and its dissociation constant from unliganded hemoglobin (C/K_{eff}) as well as on the new constraints that result from the binding of the effector (see Table 7). The lower panel of Figure 5 shows the variation in the Adair constants as a function of effector concentration. The calculations employ the Adair constants derived from the constraint values that we have found to be operative for stripped and IHP-bound hemoglobin. Characteristically, the rise in the Adair constants with effector concentration becomes significant at effector concentrations that increase in the order $K_1 < K_2 < K_3 < K_4$ and the last constant is appreciably affected only at the very largest effector concentrations ($C/K > 10^5$). One result of the orderly increase of the Adair constants with oxygenation is that the Hill coefficient at a partial oxygen pressure of 160 mm Hg becomes closer to the maximum Hill coefficient as C/K_{eff} increases (shown in the upper panel

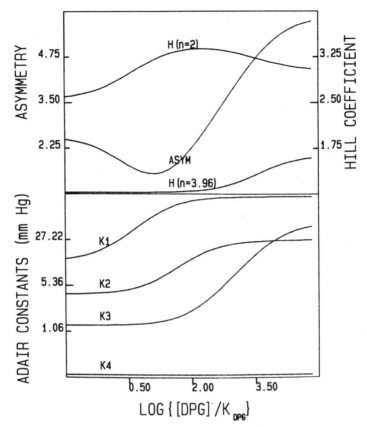

Figure 5. Upper plot: Hill coefficients and asymmetry as function of DPG/*K*(DPG). The release of the effector at high saturation permits a high degree of cooperativity to continue to near saturation. *H*(*n* = 3.6) is the Hill coefficient at 0.9 saturation. Lower plot: Progressive changes in the Adair constants with increasing DPG concentration. Note the insensitivity of K_4 to DPG.

of Figure 5); in other words as C/K_{eff} increases the cooperative effect is maintained up to degrees of ligation that progressively approach complete oxygenation. The same effect is made clear by the rise of the asymmetry with the concentration of effector, also shown in the upper panel of Figure 5.

Figure 6 shows that the effector produces a very modest decrease in the saturation achieved at ordinary atmospheric pressure (160 mm Hg) but a suitably large effect on the total oxygen unloading, assuming a constant O_2

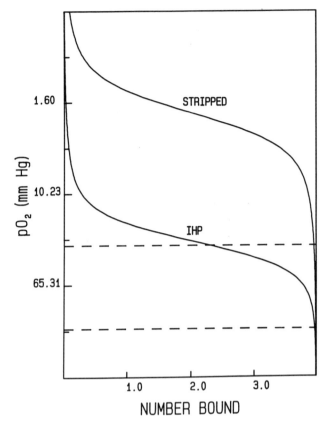

Figure 6. Effect of the decrease in O_2 affinity by IHP on the oxygen unloading in the tissues. The lower broken line corresponds to 160 mm Hg, and the upper broken line to the oxygen tension at which 60% saturation is reached under ordinary physiological conditions. Note that at 160 mm Hg virtually complete loading is achieved for both stripped and IHP hemoglobins, but efficient unloading at the tissue oxygen tension requires the presence of the effector.

pressure in the tissues. Thus one understands why a decrease in atmospheric oxygen concentration is physiologically counteracted by a seemingly paradoxical device: the decrease in oxygen affinity brought about by increased effector concentration. The increase in cooperativity and asymmetry by the effector contributes to the compensation achieved by the simple decrease in overall oxygen affinity: The larger the positive asymmetry the steeper the titration curve in the region close to oxygen saturation, and the larger the amount of oxygen released from hemoglobin for the same decrease in the partial oxygen pressure.

The asymmetry of the titration curve and the ligand distribution are closely related: Figure 7 shows the ligand distributions calculated from Eq. (**VI.2**) and the Adair constants of Tyuma *et al.* for hemoglobin, stripped (lower panel), and in the presence of 2 mM DPG (upper panel). It is

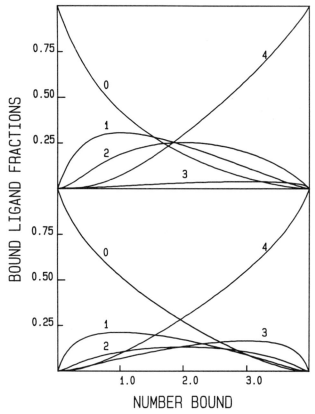

Figure 7. Ligand distributions as function of number bound for hemoglobin and hemoglobin–DPG. Lower plots are for a tetramer with Adair constants of stripped hemoglobin, and upper plots for a tetramer with those of the hemoglobin–DPG complex. Note the relative increase in the latter of molecules with a single ligand bound and the very small fraction of molecules with three ligands.

particularly noticeable that the increase in positive asymmetry reduces the fraction of molecules with three ligands bound to the point that accurate determination of K_3 from the titration data is rendered very uncertain [12]. The distribution of the oxygen molecules between the α and β chains of hemoglobin is also strongly modified by the presence of the effector. Ho and co-workers have shown by means of NMR measurements that in the presence of IHP there is a large precedence of the binding of the α over the β chain [20]. This phenomenon is directly related to the asymmetry of binding, which requires in turn asymmetry of the binding constraints. The effector provides this by the additional constraints that it places on the binding of oxygen by the β subunits through their cross-linking. Figure 8

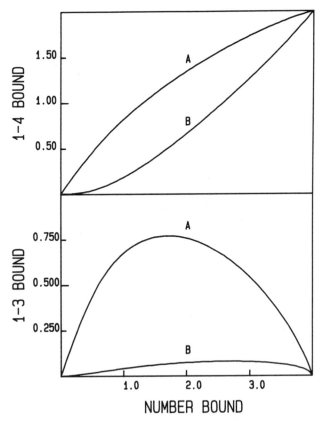

Figure 8. Plots of $n(\alpha)$, $n(\beta)$ of Eq. (**24**) for the hemo-globin–DPG complex. To be compared with the corresponding plots for stripped hemo-globin (Fig. 4).

shows plots of $n(\alpha)$ and $n(\beta)$ of Eq. (15) as a function of number bound, as deduced from sets of binding constraints that generate Adair constants and binding asymmetries characteristic of IHP-bound hemoglobin (Table 7). As a result of the large asymmetry the proportion of molecules with three ligands bound is minimal and, in consequence, in the range of physiological changes in oxygenation, the population of partially bound hemoglobin molecules has, in the main, free β and occupied α subunits.

The Bohr Effect

Following the initial observation of Bohr *et al.* [2] to the effect that carbon dioxide antagonized the oxygenation of hemoglobin, Henderson [1] explained it as due to the difference in proton dissociation constants of deoxy- and oxyhemoglobins. Proton liberation is found to be linear with oxygenation [27] indicating that it is directly linked to the state, bound or unbound, of each separate peptide chain, rather than indirectly through changes in subunit interaction. X-Ray crystallography and high-resolution proton magnetic resonance have been used in attempting to identify directly the proton-bearing amino acid residues that are responsible for the effect [28]. The original oxyhemoglobin structure suggested a ready identification at the level of the intersubunit salt bridges broken on oxygenation [9], but more recent crystallographic results (se below) cast doubt on the validity of that explanation. Ho and collaborators [28] have identified some of the protons involved by their nuclear magnetic resonances. It clearly appears that no single group is wholly responsible for the effect, but that it results from small variations in pK of a number of groups. Additionally the various contributions change considerably in their proportions when DPG or IHP is present. The conclusion is that the Bohr effect is the result of distributed effects in the conformation of the peptide chains following binding of ligands at the heme. The conclusion that the changes in conformation on oxygenation produce small changes in pK distributed over the peptide chain is borne out by a detailed calculation of the electrostatic effects expected for deoxy- and oxyhemoglobins [29].

Relations of Structure and Energetic States: Two-State Theories of Hemoglobin–Oxygen Equilibria

It has been shown above that the free energy couplings between subunit interaction and ligand binding in hemoglobin are unquestionably of first order. There must be therefore at least as many significant structures as the number of states of ligation distinguishable by their differences in the subunit constraints: There is one such state for unliganded or fully liganded hemoglobin, two for singly liganded or triply liganded hemo-globin, and four for biliganded hemoglobin—10 in all. In dealing with the energetics of binding these are reduced to five average Gibbs levels: the unliganded molecule and the four species of increasing oxygen ligation. A proposal made in 1965 by Monod *et al.* [10] envisioned hemoglobin as existing in only two significant states: low oxygen affinity (T state) and high oxygen affinity (R state). In the theory of Monod *et al.* the increase in affinity with number bound is explained by a shift in the equilibrium toward the R form with successive ligation of oxygen molecules. Two parameters are sufficient to determine the complementary fractions f_T and f_R in the two states: L, the ratio [R]/[T] for unliganded hemoglobin and c, the ratio of the dissociation constants of oxygen from R and T molecules respectively, K_R/K_T. At all intermediate states of oxygen ligation we expect the free energy of dimer association to be given by the relation:

$$\langle G \rangle = G_R f_R + G_T f_T$$

G_T and G_R are the free energies of dimer association of the T and R forms, which on the premises of the theory must correspond to those of unliganded and fully liganded hemoglobin, respectively.

In molecules of hemoglobin with two molecules of oxygen [$Hb(O_2)_2$] the theory of Monod *et al.* gives $f_R = 1/(1 + Lc^2)$ and $f_T = Lc^2/(1 + Lc^2)$. From measurements by various authors in stripped hemoglobin A, $L = 10^{-4}$; $c = 60$, which give approximately $f_R = 1/3$, $f_T = 2/3$. This is clearly incompatible with the data on the free energy of association already discussed, which show that some 90–95% of the energetic contri-bution of ligand binding to subunit association takes place on binding the first two oxygen molecules [30]. This insufficiency of the two-state theory has been recognized: Edelstein and Edsall [31] have modified the original model of Monod *et al.* to include five T states, that differ in the free energy of dimerization according to the number of molecules of oxygen bound, as well as a unique R state. Smith *et al.* [32] conclude that "either a third structural form substantially different from hemoglobin and oxy-hemoglobin exists, or the concept of a one-to-one correspondence be-tween energetic and structural states must be abandoned." The two structures originally observed by Perutz in crystals of deoxy- and oxyhemo-globin, and commonly referred by X-ray crystallographers as T and R, respectively, show substantial differences. However, Brzozowski *et al.* [33] were able to crystallize hemoglobin in which the α chains are oxygenated and the β chains are unliganded, and observe that this species has molecular characteristics that are in many respects intermediate between those of oxy- and deoxyhemoglobin. The same group of authors [34] find that hemoglobin crystallized from polyethylene glycol–water solutions can be oxygenated with changes that are small compared to those originally observed by Perutz. No appreciable displacements of the helices occur and no salt bridges are broken; the observed changes are restricted to the vicinity of the heme pockets. Thus, there is no unique crystallographic structure to which the properties of oxygenated hemoglobin in solution may be unequivocally attributed. A direct proof of the presence of two distinct species in equilibrium, T and R, that coexist in solution even in the absence of oxygen has never been given, and as already discussed in Chapter II the assumption of a plurality of protein forms in equilibrium is irrelevant for the description of the energetics of the binding of ligands by proteins. The general problem of the correspondence of the energy states with unique structural features is dealt in more detail in Chapter XVI.

References

1. Henderson, L.J. (1920) *J. Biol. Chem.* **41**, 401–430.
2. Bohr, C., Hasselbalch, K., and Krogh, A. (1904) *Skand. Arch. Physiol.* **16**, 402.
3. Hill, A.V. (1910) *J. Physiol.* **40**, iv–vii.
4. Adair, G.S. (1925) *J. Biol. Chem.* **63**, 529–545.
5. The first explicit attempt to link oxygenation to aggregation (in Lamprey hemoglobin) is due to Briehl, R.W. (1963) *J. Biol. Chem.* **238**, 2361–2365.
6. Greenwald, I. (1925) *J. Biol. Chem.* **63**, 339–349 isolated and characterized 2,3-diphosphoglycerate from dog's blood, and reported on the high concentration in the red cells. Pennell, R.B. (1964) in *The Red Blood Cell*, C. Bishop and D.M. Surgenor, eds. Academic Press, New York, gives 3.6–5 mM as the concentration of 2,3-DPG in the erythrocyte.
7. The discovery of the phosphoric esters as modifiers of oxygen affinity is due to Benesch, R., and Benesch, R.E. (1967) *Biochem. Biophys. Res. Commun.* **26**, 162–167, who also postulated the regulatory physiological action, and to Chanutin, A., and Curnish, R.R. (1967) *Arch. Biochem. Biophys.* **121**, 96–102. Benesch, R., Benesch, R.E., and Yu, C.I. (1968) *Proc. Natl. Acad. Sci. U.S.A.* **59**, 526–532 showed the reciprocity of the effects of oxygen and DPG. The increase in Hill coefficient and decrease in oxygen affinity of human hemoglobin in the presence of high phosphate concentrations preceded these observations by several years: Rossi Fanelli, A., Antonini, E., and Caputo, A. (1961) *J. Biol. Chem.* **236**, 397–400.
8. Perutz, M.F., Muirhead, H., Cox, J.M., and Goaman, L.C.G (1968) *Nature (London)* **219**, 131–139.
9. Perutz, M.F. (1970) *Nature (London)* **228**, 726–739 assigned specific functions of hemoglobin to discrete structural features. This intuitional method is

commonly employed, e.g., Royer, W., Jr, Hendrickson, W.A., and Chiancone, E. (1990) *Science* **249**, 518–521, but it does not seem reliable in view of the small energies and multiple equilibria involved.

10. Monod, J., Wyman, J., and Changeux J.-P. (1965) *J. Mol. Biol.* **12**, 88–118.

11. Benesch, R., Benesch, R.E., and Yu, C.I. (1968) Reference 7.

12. Tyuma, I., Shimizu, K., and Imai, K. (1971) *Biochem. Biophys. Res. Commun.* **43**, 423–428; **44**, 682–686. Tyuma, I., Imai, K., and Shimizu, K. (1973) *Biochemistry* **12**, 1491–1498. Mills, F.C., Johnson, M.L., and Ackers, G. (1976) *Biochemistry* **15**, 5320–5362: Extensive data on the dependence of the equilibrium on both oxygen and protein concentrations. Ackers, G.K., and Johnson, M.L. (1981) *J. Mol. Biol.* **147**, 559–582. Knowles, F.C., and Gibson, Q.H. (1976) *Anal. Biochem.* **76**, 458–486 (0.1 M phosphate buffer). Roughton, F.J.W., Otis, A.B., and Lyster, R.L.J. (1955) *Proc. R. Soc. London Ser. B* **144**, 29–54. Lyster, R.L.J., and Roughton, F.J.W. (1965) *Hvalradets Skr.* **48**, 101–196. Winslow, R.M., Swenberg, M.-L., Berger, R.L., Schrager, R.I., Luzzana, M., Samaja, M., and Rossi-Bernardi L. (1977) *J. Biol. Chem.* **252**, 2331–2337 (whole blood). Gill, S.J., Di Cera, E., Doyle, M.L., Bishop, G.A., and Robert C.H. (1987)*Biochemistry* **26**, 3995–4002. On the evaluation of the fitting procedures see: Brodersen, R., Nielsen, F., Christiansen, J., and Andersen, J. (1987) *Eur. J. Biochem.* **169**, 487–495. DiCera, E., Robert, C.H., and Gill, S.J. (1987) *Biochemistry* **26**, 4003–4008. Knowles, F.C., and Gibson, Q.H. (1976) *Anal. Biochem.* **76**, 458–486.

13. Thomas, J.O., and Edelstein, S.J. (1973) *J. Biol. Chem.* **247**, 2901–2905. Flaming, D.P., and Parkhurst, L.J. (1977) *Proc. Natl. Acad. Sci. U.S.A.* **74**, 3814–3816. Ip, S.H.C., and Ackers, G. (1976) *J. Biol. Chem.* **252**, 82–87.

14. Olson, J.S., Andersen, M.E., and Gibson, Q.H. (1971) *J. Biol. Chem.* **246**, 5919–5923 demonstrated that the rates of dissociation of one oxygen molecule from Hb$(O_2)_4$ is twofold faster than that from an isolated α chain and equal to that from a β chain, thus proving the absence of binding constraints in the oxygenated tetramer. For the oxygen affinities of the isolated chains see Brunori, M., Noble, R.W., Antonini, E., and Wyman, J. (1966) *J. Biol. Chem.* **241**, 5238–5243.

15. Mach, E. (1912) *The Science of Mechanics*. English translation of 9th edition. Open Court, Lasalle, IL, 1960, p. 425.

16. Andersen, M.E., Moffat, J.K., and Gibson, Q.H. (1971) *J. Biol. Chem.* **246**, 2796–2806 showed by kinetic and equilibrium measurements that deoxy-dimers of hemoglobin react with oxygen like the unconstrained tetramer. Likewise Mills *et al.*, ref. 12.

17. Nagel, R.L., and Gibson, Q.H. (1967) *J. Biol. Chem.* **242**, 3428–3434. See Ip and Ackers, ref. 13.

18. Brunori, M., Noble, R.W., Antonini, E., and Wyman J. (1966) *J. Biol. Chem.* **241**, 5238–5243.

19. Arnone, A. (1972) *Nature* (*London*) **237**, 146–149.

20. Viggiano, G., Ho, N.T., and Ho, C. (1979) *Proc. Natl. Acad. Sci. U.S.A.* **76**, 3673–3677.

21. Peller, L. (1982) *Nature (London)* **300**, 661–662. See also Weber, G. (1982) *Nature (London)* **300**, 603–607.

22. Desbois, A., and Banerjee, R. (1975) *J. Mol. Biol.* **92**, 479–493 found that polyvalent anions including ferricyanide and the benzene polycarboxylates have effects similar to DPG in depressing the oxygen affinity of hemoglobin.

23. Affinity labeling of hemoglobin by DPG-like ligands: Benesch, R.E., Benesch, R., Renthal, R.D., and Maeda, N. (1972) *Biochemistry* **11**, 3576–3582 (pyridoxal phosphate). Benesch, R., Benesch, R.E., Yung, S., and Edalji, R. (1975) *Biochem. Biophys. Res. Commun.* **63**, 1123–1129 (nor-2-formyl pyridoxal-5′-phosphate). Kavanaugh, M.P., Shih, D.T.B., and Jones, R.T. (1988) *Biochemistry* **27**, 1804–1808 (dithiostilbene disulfonate). The last two reagents have been shown to cross-link the β subunits.

24. Widermann, B.L., and Olson, J.S. (1975) *J. Biol. Chem.* **250**, 5273–5275.

25. Wells, R.M.G., Tetens, V., and Brittain, T. (1983) *Nature (London)* **302**, 500–502.

26. Antonini, E. (1967) *Science* **158**, 1417–1425. Antonini, E., and Brunori, M. (1970) *Annu. Rev. Biochem.* **39**, 977–1042. Guidotti, G. (1967) *J. Biol. Chem.* **242**, 3704–3712.

27. Antonini, E., Shuster, T.M., Brunori, M., and Wyman, J. (1965) *J. Biol. Chem.* **240**, 2262–2265.

28. Ho, C., and Russu, I.M. (1987) *Biochemistry* **26**, 6299–6305. Viggiano, G., Ho, N.T., and Ho, C. (1979) *Biochemistry* **18**, 5238–5247. Further on the Bohr effect: Wyman, J. (1964) *Adv. Protein. Chem.* **1964**, 223–286. Chu, A., and Ackers, G.K. (1980) *J. Biol. Chem.* **256**, 1199–2205. Busch, M.R., Mace, J.E., Ho, N.C., and Ho, C. (1991) *Biochemistry* **30**, 1865–1877.

29. Matthew, J.B., Gurd, F.R.N., Garcia Moreno, B., Flanagan, M.A., March, K.L., and Shire, S.J. (1985) *CRC Crit. Rev. Biochem.* **18**, 91–197.

30. Weber, G. (1984) *Proc. Natl. Acad. Sci. U.S.A.* **81**, 7098–7102.

31. Edelstein, S.J., and Edsall, J.T. (1986) *Proc. Natl. Acad. Sci. U.S.A.* **83**, 3796–3800.

32. Smith, F.R., Gingrich, D., Hoffman, B., and Ackers, G.K. (1987) *Proc. Natl. Acad. Sci. U.S.A.* **84**, 7089–7093.

33. Brzozowski, A., Derewenda, Z., Dodson, E., Dodson, G., Grabowski, M., Liddington, R., Skarzynski, T., and Vallely, D. (1984) *Nature (London)* **307**, 74–76.

34. Liddington, R., Derewenda, Z., Dodson, G., and Harris, D. (1988) *Nature (London)* **331**, 725–728.

IX

Equilibria Involving Covalent and Noncovalent Ligands

From the point of view of thermodynamics there is no difference in the energetics of the interactions between ligands attached by covalent bonds and those attached by noncovalent bonds, a case that was examined in detail in past chapters. The only significant dissimilarity arises because the usually much larger magnitude of the standard free energy of covalent binding is expected to result in large energies of activation for the formation and breakdown of the covalent complex and a consequent difficulty in the attainment of equilibrium. However, in the cases of physiological interest this difficulty disappears because the equilibrium with the covalently attached ligand is catalyzed by specific enzymes. In the detailed discussion of the equilibria involving ligands covalently bound to proteins it will be convenient to refer specifically to the most prevalent and biologically important of these ligands, the phosphoryl residue. However, this analysis has general value and the results are directly applicable to other cases.

Group-Transfer Potentials

Transfer of a covalently linked group from one molecule to another follows simple thermodynamic rules first pointed out by Dixon in 1951 [1], in relation to the transfer of the phosphoryl residue from one molecule to another under the influence of the specific catalysts. The rules that he derived are analogous to those valid in the transfer of protons or electrons [2] and may be deduced in the manner applicable to these latter cases. In discussing the covalent equilibria we will stress the similarities with corresponding noncovalent equilibria rather than that with those involving proton or electron transfer. Consider then the transfer of Ω from compounds $A - \Omega, B - \Omega, \ldots$ to a common acceptor $Z - R$ by way of the reactions:

$$A - \Omega + Z - R \rightarrow A - R + Z - \Omega \tag{I}$$

$$B - \Omega + Z - R \rightarrow \ldots$$

If we are interested in the transfer of an acyl group, like acetyl or phosphoryl, the choice of water as the common acceptor, that is the choice of the hydrolysis as the common transfer reaction, is a natural one. As the transfer of phosphoryl groups has the largest importance in biochemistry Φ denotes it explicitly in what follows. We would then write Reaction (**I**) as

$$A - \Phi + HOH \rightarrow A - H + \Phi - OH \tag{II}$$

The standard free energy of hydrolysis will be given by

$$\Delta G_A = \mu^0(\Phi - OH) + \mu^0(A - H) - \mu^0(A - \Phi) - \mu^0(HOH)$$

and the free energy change in this reaction, under stationary concentrations of the reagents (Chapter I), would then be

$$dG_A = \Delta G_A + RT \ln\left(\frac{\langle HOH \rangle \langle A - \Phi \rangle}{\langle A - H \rangle \langle \Phi - OH \rangle} \right) \tag{1}$$

The concentration of water would not be appreciably affected by the extent of reaction, so that $\langle HOH \rangle$ may be considered as constant. We shall take $\langle \Phi - OH \rangle$ to denote a conveniently chosen but otherwise arbitrarily fixed concentration of phosphate that will be present in every case. We may then define

$$dG'_A = \Delta G_A + RT \ln\left(\frac{\langle HOH \rangle}{\langle \Phi - OH \rangle} \right) + RT \ln\left(\frac{\langle A - \Phi \rangle}{\langle A - H \rangle} \right) \tag{2}$$

or

$$dG'_A = \Delta G'_A + RT \ln\left(\frac{\langle A - \Phi \rangle}{\langle A - H \rangle} \right) \tag{3}$$

where

$$\Delta G'_A = \Delta G_A + RT \ln\left(\frac{\langle HOH \rangle}{\langle \Phi - OH \rangle} \right) \tag{4}$$

The definition of dG'_A of Eq. (3) implies a dependence of the standard free energy change in the reaction on the concentration of phosphate chosen, a dependence appearing in Eq. (4). Notice, however, that the *difference* between any two free energies of hydrolysis remains the same if we change our standard phosphate concentration $\Phi - OH$ into another, $\Phi_1 - OH$, and thus redefine the value of dG'_A in Eq. (4). Moreover the difference in value of dG'_A measured in the two cases would equal $RT \ln(\langle \Phi_1 - OH \rangle / \langle \Phi - OH \rangle)$ and be thus easily calculable.

We define now the transfer potential of the Φ group in A, T_A, as the negative of the free energy of hydrolysis of $A - \Phi$ when in the presence of the standard phosphate concentration:

$$T_A = -dG'_A = -\left\{ \Delta G'_A + RT \ln\left(\frac{\langle A - \Phi \rangle}{\langle A - H \rangle} \right) \right\} \tag{5}$$

when $\langle A - \Phi \rangle = \langle A - H \rangle$ the equilibrium has maximal stability, the system is poised, and

$$T_A = T_A^0 - \Delta G'_A \tag{6}$$

The standard transfer potential T_A^0 is the negative of the standard free energy of hydrolysis carried out under conditions in which the phosphate concentration remains the standard one throughout the course of the reaction. As in general $dG_A < 0$ the standard transfer potential is a positive quantity, at least in most cases of interest. If various Φ carriers are arranged in a vertical scale of ascending values of T^0 it is easy to see that the vertical distance between two values gives directly the standard free energy change in the reaction in which a molar amount of Φ is transferred from the substance higher in the scale to the one below it. Thus for two Φ compounds $D - \Phi$ and $A - \Phi$, respectively,

$$T_A = T_A^0 + RT \ln\left(\frac{\langle A - \Phi \rangle}{\langle A - H \rangle} \right)$$

$$T_D = T_D^0 + RT \ln\left(\frac{\langle D - \Phi \rangle}{\langle D - H \rangle} \right) \tag{7}$$

If $\langle A - \Phi \rangle \ldots \langle D - H \rangle$ are the equilibrium concentrations $[A - \Phi] \ldots [D - H]$,

$$T_D^0 - T_A^0 = RT \ln\left(\frac{[A - \Phi][D - H]}{[D - \Phi][A - H]} \right) = -\Delta G_{DA} \tag{8}$$

where ΔG_{DA} is the standard free energy change in the reaction

$$D - \Phi + A - H \rightarrow D - H + A - \Phi \qquad \textbf{(III)}$$

A compound in the scale can preferentially transfer the Φ group to the members below it and receive it from the members above. If the T^0 value of $D - \Phi$ is $2.3RT$ (nearly 1.4 kcal at room temperature) above the value for $A - \Phi$, and we start with equal concentrations of $D - \Phi$ and $A - H$, the reaction will proceed until 90% of the Φ group is transferred from D to A. If we start with equal concentrations of $A - \Phi$ and $D - H$ the reaction will come to equilibrium when 10% of the A-linked Φ is transferred to D. If a compound X has $T^0 = 0$ and we place the form $X - H$ in contact with a phosphate solution of the standard concentration 50% will be converted to the $X - \Phi$ form. The analogy of the transfer potential scale with that for the transfer of protons in an aqueous medium is clear, but an important difference must be kept in mind. T_A^0 is a value that characterizes the couple $A - \Phi/A - H$ in the medium, not the medium itself. This situation stands in contrast to that found for protons in aqueous solutions, in which the state of ionization and pK of any compound present is sufficient to determine the pH of the solution. In Φ transfer, or more generally in all cases of group transfer, the group transfer potential is a characteristic of the couple between which the group transfer takes place (**I**), and not a general property of the medium. More explicitly, other donors besides $A - \Phi$: $B - \Phi, C - \Phi$ can generate other potentials as long as there is no spontaneous reaction in the absence of a catalyst that would ensure that the reactions involving A on one hand and B or C on the other would come to equilibrium as regards the distribution of the Φ group. The ratio $\langle A - \Phi \rangle / \langle A - H \rangle$ is an indicator of the Φ transfer ability through reactions involving this couple alone.

Modifications by Protein–Ligand Interactions

Consider a protein P that can undergo transfer of the covalently bound group Φ so that the couple $P - \Phi/P - H$ can be placed somewhere between the higher Φ donor $D - \Phi$ and the lower acceptor $A - \Phi$. From the relative places in the scale we expect that the reactions

$$D - \Phi + P - H \rightarrow D - H + P - \Phi \qquad \textbf{(IV)}$$

$$P - \Phi + A - H \rightarrow P - H + A - \Phi \qquad \textbf{(V)}$$

will take place in the directions indicated by the arrows in preference to the directions opposite to the arrows. If the protein binds preferentially a noncovalent ligand Y we expect mutual effects between Φ and Y of precisely the same nature as those that take place between two noncovalent ligands X and Y. The effects will be governed by the sign and magnitude of the free energy coupling $\delta G_{\Phi Y}$ between the covalent and noncovalent bound ligands. If $\delta G_{\Phi Y} < 0$ the form YP $-\Phi$ is stabilized with respect to P $-\Phi$. *If the magnitude of* $\delta G_{\Phi Y}$ *is larger than the difference* $T_P^0 - T_A^0$ *the prevalent direction of reaction will be inverted with respect to that shown in* (**V**):

$$\text{YP} - \text{H} + \text{A} - \Phi \rightarrow \text{YP} - \Phi + \text{A} - \text{H} \qquad (\mathbf{V'})$$

Similarly if $\delta G_{\Phi Y} > 0$ Φ will be destablized in YP $-\Phi$ with respect to P $-\Phi$, or in the language sometimes used in organic chemistry, Φ will be a better "leaving group" in YP $-\Phi$ as compared to P $-\Phi$. If $\delta G_{\Phi Y}$ is larger than the difference $T_D^0 - T_P^0$ the prevalent direction of the reaction will be opposite to that in (**IV**), namely

$$\text{YP} - \Phi + \text{D} - \text{H} \rightarrow \text{YP} - \text{H} + \text{D} - \Phi \qquad (\mathbf{IV'})$$

In this latter case the "high potential donor" D $-\Phi$ is generated from the "low potential donor" P $-\Phi$ by addition to the latter of the free energy $\delta G_{\Phi Y}$ in YP $-\Phi$ (Fig. 1) [3]. In a system in which all P is liganded in the

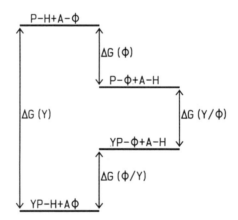

Figure 1. Scheme showing inversion of the normal sense of a covalent protein reaction by a noncovalent ligand.

form YP

$$T_{YP} = T_{YP}^0 + RT \ln\left[\frac{\langle YP - \Phi\rangle}{\langle YP - H\rangle}\right] \tag{9}$$

with

$$T_{YP}^0 = T_P^0 + \delta G_{\Phi Y} \tag{10}$$

As we discuss below Eq. (10) is the master relation governing the interconversion of chemical and osmotic energies by the agency of proteins.

Chemiosmotic Coupling across Membranes: Ion Pumping

We consider a protein in the possible forms $P - \Phi$ and $P - H$ and able to bind a noncovalent ligand Y generating two additional forms $YP - \Phi$ and $YP - H$. The possible change in the transfer potential by the addition of Y has already been described [Eq. (10)]. We suppose that these four forms of the protein are confined to a membrane M separating two chambers 1 and 2 [3]. These chambers contain the ligand Y at concentrations $\langle Y_1\rangle$ and $\langle Y_2\rangle$, respectively and characteristic Φ carriers: $D - \Phi$ with its conjugate form $D - H$ in chamber 1; $A - \Phi$ and $A - H$ in chamber 2. We assume that thermodynamic equilibrium for both Φ and Y binding is reached by the protein when in the course of its diffusion within the membrane (see Chapter X) the appropriate reactive groups come in contact with the carriers $A - \Phi$ and $D - \Phi$ and the ligand Y at either side of the membrane. The noncovalent ligand Y can traverse the membrane only when complexed with the protein as $YP - H$ or $YP - \Phi$. The transfer potentials in chambers 1 and 2 are

$$T_1 = T_D^0 + RT \ln\left(\frac{\langle D - \Phi\rangle}{\langle D - H\rangle}\right) \tag{11}$$

$$T_2 = T_A^0 + RT \ln\left(\frac{\langle A - \Phi\rangle}{\langle A - H\rangle}\right) \tag{12}$$

We consider two typical cases shown as A and B in Figure 2. In both these cases $T_{YP}^0 - T_P^0$ equals $T_D^0 - T_A^0$. In A there is cooperativity in the binding of Φ and Y to P, that is $\delta G_{\Phi Y} < 0$, while in B there is antagonism in the binding of these ligands so that $\delta G_{\Phi Y} > 0$. We further assume that in A: $T_{YP}^0 = T_A^0$, $T_P^0 = T_D^0$ and in B: $T_{YP}^0 = T_D^0$, $T_P^0 = T_A^0$. Under these assumptions the absolute value of $\delta G_{\Phi Y}$ corresponds to the difference $T_D^0 - T_A^0$

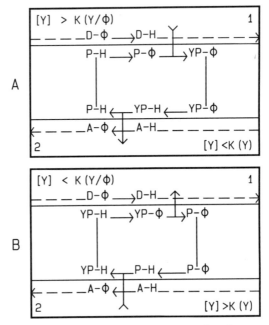

Figure 2. Scheme of free energy coupling between ion transport and synthesis of a "high energy" group donor, as discussed in text. By means of negative (A) and positive (B) free energy couplings.

in both cases. In scheme A YP $-$ Φ diffuses from the boundary of the membrane with chamber 1 to the boundary with chamber 2, while P $-$ H diffuses in the opposite direction, from chamber 2 to chamber 1 following in either case the membrane concentration gradients that are established because of the reactions at each of its boundaries. As a result there is a net transport of the ligand Y from chamber 1 to chamber 2. Starting with equal concentration [$Y_{initial}$] at both sides of the membrane, such that

$$K(Y) > [Y_{initial}] > K(Y/\Phi) \tag{13}$$

the concentration of free Y will decrease in chamber 1 and increase in chamber 2, until it falls in 1 to equal $K(Y/\Phi)$ or rises in chamber 2 to equal $K(Y)$. To keep T_1 and T_2 constant, and thus produce a uniform effect during the transport of Y, the concentration ratios $\langle D - \Phi \rangle / \langle D - H \rangle$ in chamber 1 and $\langle A - \Phi \rangle / \langle A - H \rangle$ in chamber 2 must have stationary values. These stationary values are kept by continuously introducing $D - \Phi$ into and withdrawing $D - H$ from chamber 1 while introducing $A - H$ into and withdrawing $A - \Phi$ from chamber 2 at rates determined by the transport of Y from one chamber to the other.

Conversely, in B $YP - H$ diffuses from chamber 2 to chamber 1. If at the start both chambers contain the same concentration of Y such that

$$K(Y) < [Y_{initial}] < K(Y/\Phi) \tag{14}$$

the transport of Y from 2 to 1 will in this case continue until $\langle Y \rangle$ in chamber 2 falls to equal $K(Y)$ or $\langle Y \rangle$ rises in chamber 1 to equal $K(Y/\Phi)$. Notice that in A $(\delta G_{\Phi Y} < 0)$ Y is transported from the chamber containing the Φ-carrier with the higher transfer potential to the chamber with the lower one. In B, $(\delta G_{\Phi Y} > 0)$ Y is transported from the chamber with the lower transfer potential to the chamber with the higher one. In either case the buildup of the Y gradient is achieved at the expense of the free entropy drop in the reaction

$$D - \Phi + A - H \rightarrow D - H + A - \Phi \tag{VI}$$

When 1 mol of Y has been transported across the membrane 1 mol of each of the four reagents in (VI) has been introduced or withdrawn. The total change in chemical potential $d\mu$ is

$$d\mu = \mu(D - H)_1 - \mu(D - \Phi)_1 - \mu(A - H)_2$$
$$+ \mu(A - \Phi)_2 + \mu(Y)_2 - \mu(Y)_1 \tag{15}$$

which reduces to

$$d\mu = T_2 - T_1 + RT \ln\left(\frac{\langle Y \rangle_2}{\langle Y \rangle_1}\right) \quad \text{(A)} \tag{16}$$

$$d\mu = T_2 - T_1 + RT \ln\left(\frac{\langle Y \rangle_1}{\langle Y \rangle_2}\right) \quad \text{(B)} \tag{17}$$

so that in either case we can write:

$$d\mu = T_2 - T_1 + RT \ln\left(\frac{\langle Y \rangle_H}{\langle Y \rangle_L}\right) \tag{18}$$

where $\langle Y \rangle_H$ and $\langle Y \rangle_L$ denote the high-side and low-side concentrations of Y irrespective of chamber. Under stationary concentrations of the D and A transfer couples the ratio $\langle Y \rangle_H / \langle Y \rangle_L$ will rise to a limit that is determined by the difference in the chemical potentials of the couples in the two chambers and the difference in the free energy of association of Y with $P - \Phi$ and $P - H$, that is with the absolute value of the standard free energy coupling $|\delta G_{\Phi Y}|$:

$$T_1 - T_2 \geq RT \ln\left(\frac{\langle Y \rangle_H}{\langle Y \rangle_L} \right) \leq |\delta G_{\Phi Y}| \qquad (19)$$

The limiting value of $\langle Y \rangle_H / \langle Y \rangle_L$ is determined by $T_1^0 - T_2^0$ or by $|\delta G_{\Phi Y}|$, whichever is smaller. If the smaller value is that of $T_1 - T_2$ then the whole of the free energy in Reaction (VI) is completely converted into osmotic energy but $\langle Y \rangle_H / \langle Y \rangle_L$ is below its possible maximum. If $T_1 - T_2 > |\delta G_{\Phi Y}|$ the maximum possible $\langle Y \rangle_H / \langle Y \rangle_L$ is reached but some of the free energy of transfer of Φ from D to A is not converted into osmotic energy. Complete conversion with maximum yield of ion transport will take place when the difference in standard transfer potential $T_1^0 - T_2^0$ is matched exactly by the standard free energy coupling between Φ and Y bound to the protein.

Polyvalent Protein–Ligand Complexes in Ion Pumping

These are complexes of the protein of the type $Y_N P - \Phi$ or $Y_N P - H$. If each ligand Y has an equivalent energetic effect, $|\delta G_{\Phi Y}|$ on the transfer of Φ by the protein, then the total effect on the Φ-transfer potential of $Y_N P - \Phi$ is $N|\delta G_{\Phi Y}|$. For each 1 mol of $D - \Phi$ or $A - \Phi$ transformed to the conjugate forms, $D - H$ or $A - H$, respectively, j mol of Y ($j \leq N$) is transported across the membrane. The stationary change in chemical potential $d\mu$ is related to the Y concentrations by the equation

$$d\mu = T_1 - T_2 + jRT \ln\left(\frac{\langle Y \rangle_H}{\langle Y \rangle_L} \right) \qquad (20)$$

$$T_1 - T_2 \geq jRT \ln\left(\frac{\langle Y \rangle_H}{\langle Y \rangle_L} \right) \leq |\delta G_{\Phi Y}| \qquad (21)$$

The effects are j times larger than when a single ligand Y is bound to each protein molecule [4]. As a result the Φ-transfer energy can be usefully converted into osmotic energy even when it is several times the value of $|\delta G_{\Phi Y}|$. Since N may be expected to be in the range 2–4, for a typical value of $|\delta G_{\Phi Y}|$ of 2 kcal/mol the transfer potential of ATP (8–9 kcal/mol) could be used to create an ionic gradient with considerable efficiency. Notice however that the ratio $\langle Y \rangle_H / \langle Y \rangle_L$ is still limited by $|\delta G_{\Phi Y}|$ irrespective of j, the number of ions effectively transported by one molecule of the protein. As we know from the study of the free energy coupling between ligands bound to the same protein δG_{xy} seldom exceeds 2.5 kcal; we must then expect that if our analysis is correct ionic gradients in cells and organisms will rarely show ratios $\langle Y_H \rangle / \langle Y_L \rangle$ greater than 100. This expectation is confirmed by the many available figures [5].

Reversibility of Chemiosmotic Conversion

Consider the situation present when an ionic gradient has been established and we cease to maintain the stationary concentrations of the protein forms in chambers 1 and 2. All the reactions that are depicted as occurring at the boundary of each chamber and the membrane are reversible and as a result the prevalent direction of each reaction will be inverted; the net result will be back transport of the ligand Y from the high-concentration to the low-concentration chamber. Equilibrium will be reached when the concentration of Y in both chambers becomes equal. At this point the osmotic energy provided by the initial difference in the concentration of Y would have been converted into chemical energy by increasing the difference in transfer potentials of the chambers, $T_2 - T_1$. To obtain complete reversibility of the effects we would have to introduce into and withdraw from the chambers the same reagents that were respectively withdrawn and introduced to obtain the formation of the gradient at the expense of the chemical energy in Reaction (VI). When the gradient declines we reciprocally obtain a gain in the chemical free energy of the system: The compound with the higher potential, $D - \Phi$, is synthesized by use of the low potential compound, $A - \Phi$, at the expense of the free energy stored in the ionic gradient.

Chemical Potentials and Electrochemical Potentials

Before applying the ideas sketched above to an analysis of the synthesis of ATP by ionic gradients we shall examine the possibility of changes in the chemical potential of the reactants by the presence of electric fields. Experiment shows that in the presence of an electric field the extent and character of chemical reactions may be changed. Hence we must examine whether the simple formulation of the chemical potential:

$$\mu = \mu^0 + RT \ln c \tag{22}$$

remains still valid under these conditions. We exclude from consideration those typical oxidoreductions in which the electrodes act as further reactive components of the system and confine ourselves to the effects of an "indifferent" electric field as is the field generated across biological membranes by the unequal distribution of ions across them. If the reactants and products are uncharged no changes in the concentrations of these will be generated within the field when this is applied across a restricted volume of a vessel in which reaction is taking place as shown in the scheme below:

$$
\begin{array}{c}
\hline
a \quad E = 0 \; (+) \\
\hline
b \quad E \langle \rangle 0 \\
\hline
a \quad\quad (-) \\
\hline
\end{array}
$$

We can consider the regions a and b, characterized by electric fields $E = 0$ and $E\langle\rangle0$, respectively, as two different phases and apply a general principle that refers to the free energy change in chemical reactions that take place in two phases in contact, in equilibrium with each other. Let the equilibrium constants $K(\alpha)$ and $K(\beta)$ for the reaction $A + B \rightarrow C$ in the two phases be

$$K(\alpha) = \frac{[C(\alpha)]}{[A(\alpha)][B(\alpha)]}$$

$$K(\beta) = \frac{[C(\beta)]}{[A(\beta)][B(\beta)]} \tag{23}$$

If the partition coefficients of $A \ldots C$ are $\gamma(A) \ldots \gamma(C)$ then

$$\frac{K(\alpha)}{K(\beta)} = \frac{\gamma(C)}{\gamma(A)\gamma(B)} \tag{24}$$

If the ligands are uncharged the partition coefficients will not differ from 1. If the standard chemical potentials, μ^0 were to be appreciably modified by the field, to the extent of shifting the equilibrium from its position for $E = 0$, one could obtain a continuous release of free energy from the difference in the concentration of reactants, generated because of the unequal equilibrium inside and outside the field, without actual expenditure of energy. We must then conclude that an external electric field cannot result in a change in standard chemical potential μ^0. This invariance of the standard chemical potential could also be anticipated from the small magnitude of the field strengths that are experimentally possible in comparison with those that are generated by the ions themselves: At a distance of a few Å units from an ion, the electric field is of the order of 100 abV/cm, at least 1000 times larger than the field strength across a membrane of 50 Å thickness with a typical potential difference of 0.1 V between the two sides. The entire change in chemical potential because of the presence of the electric field results only from the altered concentrations in space of those species that carry permanent charges. The concentration c then becomes a function of the coordinates and the chemical potential valid for these cases may be formulated as

$$\mu(x, y, z) = \mu^0 + RT \ln c(x, y, z) \qquad (25)$$

If the value of c in each element of space is known, no information about the electric field distribution is necessary to predict the thermodynamic equilibrium that will be obtained on reaction. Ordinarily, detailed information of the point-to-point variation of the concentration of metabolites in cells and organelles is not available and the analytical methods permit the determination of only a coarse average concentration $\langle c \rangle$ pertaining to the whole volume enclosed by a membrane. If a potential difference is established across this membrane the concentrations of ions at some distance from the membrane, $\langle c_\infty \rangle$, and close to the membrane itself, $\langle c_m \rangle$, will be appreciably different. The ratio of these concentrations is given by the Boltzmann relation

$$\frac{\langle c_m \rangle}{\langle c_\infty \rangle} = \exp\left(\frac{-dU}{kT}\right) \qquad (26)$$

where dU represents the difference in potential energy of the ions at the two locations. The partition coefficient of a charged component between the two regions that differ in field strength equals $\exp(-dU/kT)$. A pair of ions of opposite signs placed at distance r in a medium with dielectric constant ϵ has an interaction energy

$$dU = \frac{-e^2}{r\epsilon} \qquad (27)$$

where e is the electronic charge, 4.8×10^{-10} es units. For $r = 5 \times 10^{-7}$ cm, thickness of a biological membrane, and $\epsilon = 3$

$$dU = 1.5 \times 10^{-13} \text{ ergs} \qquad (28)$$

At a distance of a few Å units from the membrane the dielectric constant of water ($\epsilon = 80$) makes the electrostatic effects negligible so that we can identify dU of Eq. (**28**) with the same quantity in Eq. (**26**). At room temperature $kT = 4 \times 10^{-14}$ ergs giving

$$\frac{\langle c_m \rangle}{\langle c_\infty \rangle} \approx 10/1 \qquad (29)$$

Therefore we expect that, to a crude approximation, the concentration of the mobile ions in the immediate vicinity of the membrane differs by an order of magnitude from the concentration at some distance from it. There will be an increase in chemical potential for ions that on account of their charge preferentially accumulate close to the membrane and a similar decrease for those that accumulate away from it as a result of the electrostatic repulsion. It follows that the effect of the electric field is restricted to modification of the concentrations that exist in its absence and that the effects of the field can be duplicated *in vitro* by simply raising or lowering the concentration of the charged reagents by the appropriate amount [6]. These considerations have to be viewed against many statements in the literature on bioenergetics implying that the electric field plays a role in the chemistry that is not restricted to a change in the concentration of the reagents, but involves a direct effect of the field on the chemical reactions. In our opinion neither an experimental proof nor convincing theoretical arguments have been advanced in favor of such claims. The real importance of the potential differences across membranes is to be found in the modifications of the distribution of ions and charged molecules in the immediate vicinity of the membrane, a circumstance that may be all important for some metabolic process.

The Synthesis of ATP through Ionic Gradients

The synthesis of adenosine triphosphate (ATP) as a result of oxidoreductions occurring during aerobic metabolism was discovered in the middle 1930s and posed a problem that is still not completely solved in spite of much effort. ATP possesses one of the highest transfer potentials of the phosphoric esters present in organisms and is the universal phosphorylating agent that permits the synthesis of a large variety of compounds with lower potentials by simple transfer reactions catalyzed by specific phosphokinases. It was originally thought that an unstable higher potential phosphoryl donor was generated in oxidative metabolism and reacted rapidly to phosphorylate ADP, in analogy with the generation of ATP from ADP in glycolysis, but a search of many years failed to detect the elusive phosphorylated intermediate. In the Na^+–K^+-ATPase and Ca^{2+}-

ATPase it is clear that the physiological function of these enzymes is the generation of the respective ionic gradients by means of the free energy released in the ATP hydrolysis, while the demonstrable ATP synthesis by the declining gradients is simply a consequence of the reciprocity of the energetic coupling [7]. A net synthesis of ATP by the osmotic energy requires the formation of the gradient through some additional free energy source. Mitchell [8] proposed that oxidoreductions resulted in an unequal distribution of protons at both sides of the mitochondrial membranes, thus providing the necessary ionic gradient. This proposal has been generally accepted, though the experimental evidence is still rather indirect [6], and the exact place and time of the proton intervention in promoting the ATP synthesis are not yet known. We have already discussed the limited role of the electrostatic gradients in these processes. It is common to find in discussions of the generation of ATP through ionic gradients in the literature, equations that include the electrical potential across the membrane as a term additional to the one represented by the difference in ion concentrations. What is not provided is a description of the mechanisms that couple the decline of the electric field and the generation of additional chemical energy, as indispensably required by free energy conservation. On the other hand the coupling of the free energies of binding of covalent and noncovalent ligands offers a sufficient first basis for understanding the energetics of the synthesis of ATP by the osmotic energy of ionic gradients. A major consequence of this assumption is that the generation of ATP from ADP and inorganic phosphate should proceed in the absence of membranes, simply by subsequent exposure of the ATPase to solutions with the ionic compositions that are present at the two sides of the membrane. The study of the Na^+–K^+-ATPase and Ca^{2+}-ATPase [7] has provided abundant experimental evidence for this statement.

Microscopic Implications of the Chemiosmotic Conversion by Free Energy Couplings between Bound Ligands

We would like to know how thermodynamic necessity manifests itself at the level of the discrete interactions between the protein and the bound ligands. Clearly a thermodynamic analysis cannot provide us directly with the operative microscopic mechanisms, but it is useful in ruling out some of the possibilities and in drawing our attention to the steps that may be significant. For the purpose of such an analysis it is convenient to consider the enzyme as a "black box," E inside which the reaction

$$ATP + HOH \leftrightarrow ADP + P_i$$

takes place. Within the enzyme box this reaction has a characteristic equilibrium constant K_E:

$$K_E = \frac{[ATP]_E[HOH]_E}{[ADP]_E[P_i]_E} \tag{30}$$

that differs considerably from that determined by the standard free energy change of hydrolysis of ATP [9]. The concentrations that appear in (30) are those of the complexes of the enzyme with the four reagents. By this we mean that on a microscopic examination of 1 mol of enzyme–substrate complexes the substrates present in those complexes are represented by the quantities $[ATP]_E \ldots [P_i]_E$. In the solvent we expect an equilibrium of the same components with constant K_S the meaning of which we shall discuss later; for the present we need only to assert that it corresponds to an equation like (30) in which the concentrations of the reagents, left unsubscripted, and those prevalent in the solvent after the enzyme has brought the system to a state of apparent equilibrium, as operationally defined in Chapter II. The resultant differences between the chemical potentials of the products and reactants in the solvent, ΔG_S, and in the enzyme box, ΔG_E, are defined by the equations

$$\Delta G_S = \mu(ADP) + \mu(P_i) - \mu(ATP) - \mu(HOH)$$

$$\Delta G_E = \mu(ADP_E) + \mu(P_{iE}) - \mu(ATP_E) - \mu(HOH_E) \tag{31}$$

where the unsubscripted reagents refer to those in the solvent. The differences in the chemical potentials of the reagents $\mu(ADP) - \mu(ADP_E), \ldots, \mu(HOH) - \mu(HOH_E)$ are, respectively, their free energies of transport from the solvent to the enzyme box, $dT(ADP) \ldots dT(HOH)$. It follows that

$$\Delta G_S - \Delta G_E = dT(ADP) + dT(P_i) - dT(ATP) - dT(HOH) \tag{32}$$

We consider that to a first approximation changes in the solvent composition leave ΔG_E unchanged. The value of ΔG_S, and therefore the equilibrium reached in the solvent, is determined by both ΔG_E and the free energies of transport of the reagents from solvent to enzyme according to Eq. (32). De Meis [10] has demonstrated experimentally that the addition of glycerol, dimethyl sulfoxide, or polyethylene glycol results in changes in the equilibrium (K_S) reached in the enzymic hydrolysis of pyrophosphate. The new equilibrium is characterized by the presence of substantial amounts of pyrophosphate so that reaction in the absence of initial pyrophosphate leads to its synthesis by pyrophosphatase. We expect this effect to follow the decrease of the free energy of transport of water from enzyme to solvent. In the same way the existence of an antagonistic free energy coupling of ATP with another ligand bound to the protein may be regarded as leading to a decrease in the free energy of transport of ATP from enzyme to water, thus facilitating its synthesis. Considered from this point of view an explanation of ATP synthesis involves giving answers to two questions:

1. What are the intimate mechanisms that reduce the free energy of hydrolysis of ATP, *in the enzyme*, from a value of several kcal mol^{-1} to one that may not differ greatly from zero.
2. What are the influences that change the values of the free energies of transport of the reagents and thus facilitate the release of ATP to the solvent.

We may note at this point that we have not assumed in this discussion that there is an absolute value of K_S that is independent of the catalyzed reaction. ATP, ADP, and P_i in water solutions are metastable compounds that will persist unchanged for a time sufficiently long to consider that their equilibrium is now determined by K_E and the free energies of transport of the reagents according to Eq. (32). We note, however, that for K_S to assume an arbitrary value that differs from that of the standard free energy of hydrolysis of ATP it is necessary to introduce further reagents (bound ligands, additional solutes), the role of which is to couple the hydrolysis to the processes of transport of the reagents from the enzyme box to the solvent. We are thus able to understand how the same enzyme can catalyze the equilibration of ADP and P_i with ATP with very different results: These depend on the values of the free energies of transport of the reagents from enzyme to solvent. Evidently, this free energy of transport largely results from differences in the free energy of solvation of the substrates in enzyme and solvent, a factor that has been much emphasized. The homogenization of mitochondria in a volume of solvent large in comparison with that of the intact organelle will have the effect of increasing K_S through the change in the free energy of transport of ATP to the solvent, thus decreasing the yield of ATP synthesis. It is understandable that such an observation could be wrongly interpreted as indicating that some indispensable cofactor or a diffusible, unstable phosphorylating agent had been lost in the process. The presence of high concentrations of enzymes that utilize, and thus preferentially bind, ATP will further promote the transfer of ATP to water and result in a value of ΔG_S that approaches that of ΔG_E. In this way the uptake of ATP for its further utilization in metabolism will increase the yield of ATP synthesis, and by tightly coupling ATP synthesis and utilization will permit the regulation of the former by the latter simply through the law of mass action.

References

1. Dixon, M. (1951) *Multienzyme Systems*. Cambridge University Press, London.
2. Clark, W.M. (1972) *Oxidation-Reduction Potentials of Organic Systems*. R.E. Krieger, Huntington, NY.
3. Weber, G. (1972) *Proc. Natl. Acad. Sci. U.S.A.* **69**, 3000–3005.
4. Weber, G. (1974) *Ann. N.Y. Acad. Sci.* **227**, 486–496.
5. For a discussion of the physiological significance and the magnitude of ionic gradients see Katz, B. (1966) *Nerve, Muscle and Synapsis*, Chapter 4. McGraw-Hill, New York.
6. Jagendorf, A., and Uribe, E. (1966) *Proc. Natl. Acad. Sci. U.S.A.* **55**, 170–177.
7. Knowles, A.F., and Racker, E. (1975) *J. Biol. Chem.* **250**, 1949–1951. De Meis, L., and Vianna, A.L. (1979) *Annu. Rev. Biochem.* **48**, 275–292.
8. Mitchell, P. (1961) *Nature (London)* **191**, 144–148 is the original statement of chemiosmotic coupling. A detailed exposition of "vectorial metabolism" and its relations to transport is given in Mitchell, P. (1976) *Biochem. Soc. Trans.* **4**, 399–430. For a review of the arguments involved in this subject see Fillingame, R.H. (1980) *Annu. Rev. Biochem.* **49**, 1070–1113.
9. Knowles, J.R. (1980) *Annu. Rev. Biochem.* **49**, 877–919. An example of the importance of the free energy of transport in determining the course of enzymic reactions: Eggers, D.K., and Blanch, H.W. (1988) *Bioprocess. Eng.* **3**, 83–91.
10. De Meis, L. (1989) *Biochim. Biophys. Acta* **973**, 333–349.

X

Intramolecular and Solvent Interactions of Proteins

Intramolecular Interactions in Proteins

Knowledge of the interactions among the component parts of the protein molecule can furnish the qualitative explanation of many experimental facts concerning their stability, their solubility in water, their ability to bind ligands, and their possible transfer to an immiscible phase. In principle, it should also include a quantitative, and therefore predictive, account of the internal interactions of the peptide chain that are responsible for the specific protein structure. To examine the main types of interactions within and between proteins, as regards their relative strength and their stability vis-à-vis the thermal agitation, some elementary models must be used. In employing these models to interpret the interactions responsible for the structural characteristics of proteins, one must remember the additional complexities present in their case, which are absent in simpler systems. These include the following:

1. Interactions of very different type.
2. Small contribution to the total from any one of them.
3. Dependence on exact geometry of the interacting parts.
4. The existence of an important internal dynamics that arises from the small values of the many elementary interactions and their individual instability toward the thermal agitation.

Electrostatic Interactions due to Fixed Charges

Essentially the molecular interactions are always of electrostatic character. The simplest one is the Coulombic interaction between two charges of magnitudes q_1 and q_2 placed at distance r in a medium of dielectric

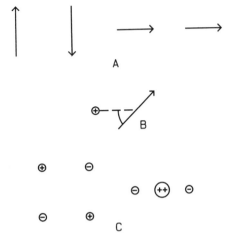

Figure 1. (A) Dipole interactions: sequential and parallel. (B) Charge–dipole interaction: angular dependence. (C) Quadrupole.

constant ϵ. The potential energy of the system equals $u(r)$, the work done in bringing the charges from practically infinite distance to r:

$$u(r) = \frac{-q_1 q_2}{\epsilon r} \tag{1}$$

If $q_1 = q_2 = e$, the electronic charge ($= 4.8 \times 10^{-10}$ esu), at a temperature of 298 K (25° C), in vacuum ($\epsilon = 1$), the distance at which $u(r)$ equals the thermal energy kT, is $r_{kT} = 557$ Å. If $\epsilon = 80$ (water) $r_{kT} = 7$ Å. These figures are indicative of the damping of the electrical interactions by a polar solvent. The simplest fixed distribution of charges that preserves total zero charge is the dipole consisting of a pair of charges $+q$ and $-q$ separated by a distance $2l$. A single ionic charge of magnitude e placed at distance r from the midpoint of the dipole (Fig. 1a) is subjected to energies of coulombic attraction and repulsion by the two monopoles of the dipole of $-eq/[\epsilon(r - l)]$ and $eq/[(\epsilon(r + l)]$, respectively. Therefore

$$u(r) = -\frac{eq}{\epsilon}\left(\frac{1}{r - l} + \frac{1}{r + l}\right) = \frac{-2lqe}{\epsilon(r^2 - l^2)} \tag{2}$$

$2lq$, the product of charge and charge separation, is m, the electric moment of the dipole, so that the last equation is conveniently written as $u(r) = me/[\epsilon(r^2 - l^2)]$. If $l \ll r$ the interaction is approximated by the simpler relation $u(r) = -me/(\epsilon r^2)$. This approximation implies that the dipole charges are close enough to constitute a "point dipole" of negligible dimensions and strength m. The point dipole approximation underestimates the attraction by a fraction $(l/r)^2$. If the external charge is at the minimum distance between independent atoms (i.e., the sum of the van der Walls radii, ~ 3.1 Å) and $2l$ is the covalent bond distance (~ 1.5 Å) then $r = 4.25$ Å and $(l/r)^2 = 0.11$. This is then the practical limit of the relative error associated with the point dipole approximation, when used to compute the interaction between an ion and a small dipole. Dipole moments are most often quoted in Debye units, $D = 10^{-18}$ esu units. In the point dipole approximation, the interaction energy of a dipole with $m = 3D$ with an electronic charge $q = e$ gives $r_{kT} = 17$ Å for $\epsilon = 1$ and < 1 Å for $\epsilon = 80$. Evidently the assumption that the solvent acts as an intervening medium with macroscopic dielectric constant ϵ breaks down when r is so small that dipole and ion are in virtual contact. In these cases the computation of the interactions must include those with the additional sources of charge in the solvent or other environment, with ϵ set equal to unity. Additionally the in-line orientation of the charges, assumed in the derivation of Eq. (2), can be expected to hold only when the electrostatic energy is much larger than the thermal energy, as the unique orientation of ion and dipole will only then be permanent. If dipole and charge are not in line but the direction of the dipole makes an angle θ with the line from dipole center to charge q,

$$u(r, \theta) = -u_0 \cos\theta \qquad (3)$$

where $u_0 = mq/(r^2 - l^2)$. If $\theta > \pi/2$ attraction is changed into repulsion. The average value of $u(r, \theta)$ depends on the distribution of orientations: With unrestricted orientations their statistical distribution is that worked out in 1906 by Langevin for magnetic dipoles placed in an external magnetic field: The relative weight of each orientation is given by the Boltzmann factor $\exp(u_0 \cos\theta/kT)$, and the mean value, $\langle \cos\theta \rangle$, obtained by integration is

$$\langle \cos\theta \rangle = \text{ctanh}(u_0/kT) - kT/u_0 \qquad (4)$$

By developing $\text{ctanh}(u_0/kT)$ in powers of u_0/kT, retaining only the square term

$$\text{if } \frac{u_0}{kT} \ll 1 \quad \text{then} \quad u \to \frac{u_0}{3kT}$$

$$\text{if } \frac{u_0}{kT} > 1 \quad \text{then} \quad u \to 1 - \frac{kT}{u_0} \qquad (5)$$

Equations (5) show that in the Langevin distribution the *average* energy of interaction increases more slowly with the energy of best interaction (u_0) than the simple Boltzmann average, because of the spatial restrictions to the most favorable orientation.

The interaction of two dipoles *in the most favorable orientation* (Fig. 1A, right pair), given by the sum of the attractive and repulsive terms is

$$u(r) = \frac{-2m^2}{r(r^2 - l^2)} \tag{6}$$

or in the point dipole approximation

$$u(r) = \frac{-2m^2}{r^3} \tag{7}$$

The next more complex distribution of charges with total charge zero is the quadrupole (Fig. 1C). In the point dipole approximation its interaction with an ion has energy $u(r) = -2me/r^3$. The form is similar to that for the ion–dipole case but in the ion–quadrupole case the quadrupole field decays with r^{-3} while it does with r^{-2} in the case of the dipole field. This illustrates a general property of the neutral distributions of charge. The larger the number of charges participating, the steeper the decline of the electric effects with the distance. Globular proteins have a very symmetric distribution of charges and the small attractive effects on similar molecules at the isoelectric point are important in promoting their solubility in water.

Induced Molecular Dipoles

All molecules are polarizable. When placed in an electric field a separation of charge occurs with appearance of an electric moment of strength m_{ind}. The field strength, that is the force exerted on unit charge by another charge q placed at distance r is $F = q/r^2$. The moment induced by this field is

$$m_{ind} = \alpha F \tag{8}$$

and the proportionality constant α is the polarizability of the molecule. The dimensions of m are charge $\cdot$ cm and that of F is charge cm^{-2}; therefore α has dimensions of cm^3, a volume. Atomic polarizabilities are of the order of 1–3 ml mol^{-1}. Molecular polarizabilities (e.g., naphthalene) can reach 100 ml mol^{-1} and a is therefore of the order of 10^{-23} ml/molecule. The work of induction, that is the energy required to produce m_{ind}, is given by

$$u_{ind} = \int_0^l Fe\,dl \tag{9}$$

where e is the electronic charge. Replacing F by $m_{ind}/\alpha = el/\alpha$ and performing the integration gives

$$u_{ind} = \frac{m_{ind}^2}{2\alpha} = \frac{\alpha F^2}{2} \tag{10}$$

if the inducing field is from an ionic charge then $F = e/r^2$; if from a permanent dipole $F = m/r^3$ and then

$$u_{ind}(\text{charge}) = \frac{\alpha e^2}{2r^4}; \qquad u_{ind}(\text{dipole}) = \frac{\alpha m^2}{2r^6} \tag{11}$$

Comparison of Eq. (11) with Eqs. (2) and (7) shows that the induction effects decay with the distance still more steeply than those owing to distributed permanent charges. In estimating the inductive effects of a dissolved ion on a molecule it must be kept in mind that

1. Ions require a strong polar solvent to separate them from others of opposite charge and the solvation increases the distance of closest approach of ion and molecule beyond that of their van der Walls radii.

2. The interposed polar molecules screen the electric effects of the ion in proportion to the macroscopic dielectric constant of the solvent. When the solvent is water these two effects combine to decrease the strength of the interactions by about two orders of magnitude. If $r = 3.1$ Å and the molar polarizability, $N\alpha = 6$ ml mol^{-1} the last equation would give $u_{ind} = 30\ kT$ at room temperature. Screening and separation by the solvent would reduce this to about $0.3\ kT$.

Mutual Inductive Effects: London–van der Waals Forces

In the absence of any permanent dipole moment molecules and atoms show mutual inductive effects. Momentary departures from coincidence of the centers of mass of positive and negative charges in an atom or molecule result in the appearance of coupled oscillations of the electrons in their near neighbors. The attraction resulting from these oscillations, first described by London in 1929, is responsible for the liquefaction of gases. Their dependence of the sixth power of the distance [second Eq. (11)] explains the dependence of the condensation of the permanent gases on the inverse square of the volume postulated by the van der Waals equation, hence their usual designation as London–van der Waals forces. These electronic oscillations are of the same kind as those elicited by an oscillating electric field of the appropriate frequency and are thus responsible for the dispersion of light, hence the name of "dispersion forces" commonly applied to them; the term "van der Waals forces" is usually meant to comprise all the forces among neutral molecules [1]. We may simply write the distance dependence of the dispersion forces as $u(r) = -b/r^6$. For a couple of oscillators consisting of the electronic mass elastically bound to an infinite mass $b = 3\alpha N h \nu_0$. The molar energy of the atomic oscillator, $N h \nu_0$, is equated with the ionization energy of molecules (~ 150 kcal/mol) on the assumption that the amplitude of the oscillations at the ionization point would be much larger, but the frequency is substantially the same as that involved in the coupling. This gives $h \nu_0 \sim 10^{-11}$ ergs. At a distance of 3.2 Å a pair of atoms with molar polarizabilities of 1 ml would have $u(r) = 2.5 \times 10^{-14}$ or $kT/2$ at room temperatures. London [2] pointed out that these forces are additive over a collection of molecules and are thus responsible for the coherence of liquids and molecular solids. Unlike the effects caused by permanent charge interactions, or induced by the permanent charge of a second molecule, a pair of molecules interacting solely by dispersion forces need not be disturbed by the presence of a third one as occurs in the former cases. The strong dependence of the London–van der Waals forces on the distance indicates that the attractive effects from this source on a bound ligand is limited to the atoms in the protein in direct contact with it and this circumstance furnishes the basis for the requirement of close fitting between protein and ligand.

Complete Interaction Potentials of Molecules

At very short distances the van der Waals attraction between neighboring atoms is replaced by the repulsion caused by the overlapping of the electron shells. To account for the dependence of both attraction and repulsion with distance the interaction energy of a pair of atoms may be written as first done by Born:

$$u(r) = k_1 r^{-s} - k_2 r^{-t} \tag{12}$$

where the first term on the right-hand side describes the short-range electrostatic repulsion between complete electron shells. From various experimental data s is known to have a value between 8 to 12. The second term denotes the attraction resulting from any of the various forces described above. The classical description of the interatomic potential caused by the London–van der Waals forces, is that proposed by Lennard Jones [3]:

$$u(r) = Ar^{-12} - Br^{-6} \qquad (12')$$

The minimum in the interaction energy occurs at distance r_0. This is found by setting $du(r)/dr = 0$, which gives

$$-12 Ar_0^{13} + 6 Br_0^7 = 0; \qquad A = Br_0^6/2; \qquad u(r_0) = -B^2/(4A) \quad (13)$$

replacing A and B in (12) by their values in the last equations

$$u(r) = u(r_0)\left[(r_0/r)^{12} - 2(r_0/r)^6\right] \qquad (14)$$

Setting $y = (r_0/r)^6$ the last equation yields the quadratic

$$y^2 - 2y + \left[u(r)/u(r_0)\right] = 0 \qquad (15)$$

The two values of r_0/r for which $u(r)/u(r_0) = 1/2$ are 1.093 and 0.815. Thus a decrease of 50% of van der Waals energy is brought about by an increase of 18% or a decrease of 9% of the interatomic distance. If $r_0 = 3.1$ Å these amount to $+0.6$ and -0.3 Å, respectively. These figures give an indication of the high precision required of the atomic coordinates to permit a direct reliable estimation of interaction energies in proteins starting from the X-ray structural data.

To obtain the complete interaction potentials between molecules we must add to the shortest range interactions described by Eq. (12) the electrostatic effects resulting from the presence of charges. It is convenient to assign monopole charges q_i to the atoms in the structure and to compute and add all the monopole–monopole interactions to those described by the Lennard–Jones potential. In this way the complete potential takes the following form:

$$u(r) = Ar^{-12} - Br^{-6} + \sum_{i\langle\rangle j} \frac{q_i q_j}{r_{ij}} \qquad (16)$$

van der Waals potentials of the form of Eq. (16) are referred to as 12-6-1 potentials from the inverse powers of the distances entering in it. Calculations have shown that a 9-6-1 potential often gives similar or even better agreement with experimental data.

In attempting to apply the potential functions of interaction between elementary structures to something as complicated as a relatively small molecule in solution, let alone to proteins and other macromolecules, many limitations have to be kept in mind:

1. The additivity of the effects due to several independent causes, expressed in equations such as (16), is not at all guaranteed.

2. Even in molecules of small size in solution the changing interactions of each of its atoms with those of the solvent and the concomitant fluctuations in energy have prevented up to now the *ab initio* computation of their solution properties.

3. The summation of a relatively large number of elementary interactions of opposite sign leads to a final result that is very much smaller than the sum of the absolute values of the component terms. In consequence, the accumulated errors deprive of significance the coincidence of computed and experimental values. In fact, agreement with the experiment may be always enforced by minimal changes in the various parameters (charges, distances, polarizabilities) that determine the outcome of the computation. From these observations we have to conclude that the present day knowledge of the elementary atomic interactions is insufficient to permit the exact computation of the thermodynamics or kinetics of molecular interactions in solution. However, they do provide an estimate of the magnitude of molecular interactions and therefore of the qualitative limits for the interpretation of the functions of proteins in terms of their detailed atomic structure. Some degree of success has been achieved in calculations of the purely electrostatic energies of proteins in applications to problems such as the pK of proton-bearing dissociating groups and the long-range interactions between protein parts [4].

Solubility of Proteins in Electrolyte Solvents

The insolubility of globular proteins in media of low dielectric constant follows from the presence of charges residing at the acidic and basic amino acids. The former are from the glutamic and aspartic residues, and the latter from histidine, lysine, and arginine. Because of the equal numbers of positive and negative charges, the proteins molecules at the isoelectric point are subject to mutual attraction. The intermolecular distance at which this attraction is just sufficient to overcome the thermal agitation depends on the distribution of charges in the molecule and the dielectric constant of the solvent. The preponderance of charges of opposite signs at antipode positions in the molecule increases the dipole moment of the protein and favors the molecular association, while a very uniform distribution of opposing charges over the surface of the molecule favors a high solubility at the isoelectric point in media of low ionic strength. From the simple fact that one can prepare solutions containing 0.5 g of serum albumin or hemoglobin per ml of water one can deduce that these molecules must have a very uniform distribution of charge [5] and therefore a relatively small dipole moment. This supposition is verified by direct measurements: The dipole moments of serum albumin and hemoglobin are 450 and 380 Debye units, respectively [6]. The short-range neutralization of charges by others of opposite sign at nearby places is clearly demonstrated by the fact that although there are over 100 positive charges and an equal number of negative charges distributed over the surface of these proteins, the dielectric increment effects produced by addition of protein to solvent are those that would result if the molecules carried 1–2 electronic charges at antipode positions (55 Å). The effects of changes in the dielectric constant of the solvent are demonstrated by the precipitating action of the freely miscible alcohols (methanol, ethanol, and propanol) when added to aqueous solutions of proteins at the isoelectric point. Away from the isoelectric point the electrostatic attraction is replaced by repulsion, and increased solubility, and this effect increases as the dielectric constant of the solvent decreases. The differential effects of pH on the solubility of proteins are most marked in their solutions in alcohol–water mixtures and this circumstance has been used to advantage in the purification of the plasma proteins by ethanol precipitation at low temperature [7].

The uniform distribution of charge is not found in every protein molecule and is in fact rather characteristic of those molecular species that are found in high concentration in their natural environments. Some blood globulins and insulin are, for practical purposes, insoluble at the isoelectric point because of their relatively much higher dipole moment. In these cases solubility is increased by the addition of neutral salts, a phenomenon observed early in the study of proteins and designated as "salting in" [8]. Evidently, salting in results from the screening of the molecular dipoles by ionic charges of opposite signs. Equivalently one may say that the monopole charges in each separate molecule are replaced by dipoles, formed by attraction of the counterion, and in this way the distance of closest approach without formation of a lattice of molecular complexes is decreased. In contradistinction, the addition of high concentrations of neutral salts (salting out) results in precipitation of most proteins [9]. This property was the basis of the earliest method of isolation and characterization of proteins. Its exact mechanism of action is still obscure: we may surmise that at these high concentrations every protein charge in contact with solvent is neutralized and the resulting dipoles distributed over the protein surface result in sufficient intermolecular attraction to greatly reduce the solubility. Additionally the high salt concentration may favor reduction of the hydration layer immediate to the protein, permitting a closer approach and stronger electrostatic attraction between the molecules.

References

1. For review of typical molecular interactions see Hirschfelder, J.O. (1965) In *Molecular Biophysics*, B. Pullman and M. Weissbluth, eds. Academic Press, New York, pp. 325–342.

2. London, F. (1936) *Trans. Faraday Soc.* **33**, 8–16 is the classical paper on interactions of neutral molecules. The importance of the London forces in the close fitting of proteins and ligands was first stated by Pauling, L. (1948) *Nature (London)* **161**, 707–709.

3. Fowler, R. (1955) *Statistical Mechanics*, Chapter X. Cambridge University Press, London. On application of complete potentials to molecular interaction calculations: Lifson, S., Hagler, A., and Dauber, P. (1979) *J. Am. Chem. Soc.* **101**, 5111–5121.

4. Honig, B., and Sharp, K.A. (1990) *Annu. Rev. Biophys. Biophys. Chem.* **19**, 301–332.

5. Wada, A., and Nakamura, H. (1981) *Nature (London)* **293**, 757–758.

6. Oncley, J.L. (1943). In *Proteins, Aminoacids and Peptides*, Chapter 22. E.J. Cohn and J.T. Edsall, eds. Reinhold, New York, pp. 543–567. Cohn and Edsall's classic book, written in 1942, is still the most authoritative source on the bulk properties of proteins in solution.

7. Cohn, E.J., Strong, L.E., Hughes, W.L., Jr, Mulford, H.J., Ashworth, J.N., Melin, M., and Taylor, H.L. (1946) *J. Am. Chem. Soc.* **68**, 459–475. The fractionation of plasma proteins by ethanol at low temperature is the only described attempt to conservatively separate and characterize *all* the proteins of an animal tissue.

8. Green, A.A., and Hughes, W.L. (1951) *Methods Enzymol.* **1**, 67–90. A paper that should be read by anybody involved in protein purification.

9. Scatchard, G. (1943) In *Proteins, Aminoacids and Peptides*, E.J. Cohn and J.T. Edsall, eds. Reinhold, New York, pp. 20–72.

XI

Transfer of Proteins to Apolar Media and the Dynamic Interactions of Proteins and Membranes

Partition of Molecules between Water and an Immiscible Phase

In the transfer of a molecule between two immiscible phases bonds between one solvent and the molecule must be broken and the energy required for this process is to be partially or totally paid by the interactions of the dissolved molecule with those of the second phase. The difference between the energies demanded by these two processes will determine the partition coefficient of the solute between the two phases.

The molecules of an ideal apolar solvent attract each other exclusively by dispersion forces. They interact with an apolar solute molecule by forces of the same character and comparable strength and as a result the apolar solvent is unable to reject an apolar molecule that attempts to penetrate it. A different state of affairs exist in a solvent of which the molecules are permanent dipoles, which interact with each other through forces comparatively much larger than the London forces. Penetration of a foreign molecule into the solvent will be resisted unless the bonds to be formed between solute and solvent are of strength comparable to those between the solvent molecules, and this will evidently require a polar solute. The rejection of a nonpolar solute by a polar solvent has been designated a "solvophobic" (in the case of water a "hydrophobic") effect. As water is a solvent with one of the highest dipole moments, and the highest dipole density, the solvophobic effects are particularly strong in its case. From a macroscopic point of view the molecular interactions that characterize a liquid and its ability to mix with other liquids is related to the so-called "internal pressure" [1]. The concept of an internal pressure is derived in a straightforward manner from the statement of the first law of thermodynamics discussed in Chapter I.

$$dU = dq - p\,dV \tag{1}$$

where dU is the change in the internal energy of the system, dq the change in heat content, and dV that of volume. Substituting dq by $T\,dS$, the change in internal energy with respect to volume at constant temperature is

$$\left(\frac{dU}{dV}\right)_T = T\left(\frac{dS}{dV}\right)_T - p \tag{2}$$

Introducing in Eq. (2) the Maxwell relation $(dS/dV)_T = (dp/dT)_V$ the last equation becomes

$$\left(\frac{dU}{dV}\right)_T = T\left(\frac{dp}{dT}\right)_V - p \tag{3}$$

As U has dimensions force x length and V dimensions are area x length, dU/dV has those of force/area = pressure, and provides a measure of the cohesive forces within the liquid. $(dp/dT)_V$ can be equated with the ratio α/β where α is the coefficient of thermal expansion and β the compressibility coefficient of the liquid. If this ratio is given in bars per degree it is found to vary in different liquids from about 7 to 40, the lower values corresponding to apolar liquids like hexane and the highest to water. The corresponding internal pressures at room temperature range then from 2000 to 12,000 atmospheres. Evidently the contribution of p in Eq. (3) can be neglected when this is as low as 1 bar. The internal pressure is determined by the nature of the forces that bind the molecules of the liquid and introduces a scale that permits visualization of the difficulty of penetration of polar liquids by apolar molecules, and also explains the decrease in volume of apolar molecules introduced in a polar medium.

The time-dependent interactions among the molecules in liquid water have been computed by Rahman and Stillinger, employing the methods of molecular dynamics [2]. They assumed that the interactions were exclusively of two sources: the London forces and those between monopoles. As neon and water are isoelectronic, the experimental van der Waals parameters for the interaction between neon atoms were employed in the calculations. Monopole charges, attached to the positions of the hydrogen and oxygen, were derived from the known dipole moment of water. With these simple ingredients they obtained a reasonable approximation to the equilibrium properties of liquid water and showed that many complex structural regularities, including the hydrogen bonds, arise naturally in the computations [2] as a joint result of the dynamics and the electrical interactions. In many topics of biological interest it has been customary in the past to introduce "the structure of water" as an important determinant of the properties of solutions, particularly those of proteins. It is the great merit of the computations of Rahman and Stillinger to have laid to rest such a vague and unhelpful concept: Whatever structural characteristics may be attributed to water, the determinant of the energetics and dynamics of water solutions need not be sought in other than the simple dipolar and dispersion interactions employed in their calculations.

In discussions of the properties of water as a solvent for partially or wholly apolar molecules it is common to attribute the observed effects to the solvophobic property, while the possibilities of water as a potentially apolar solvent are disregarded. However, these are always present: The interactions between the water dipoles result in self-neutralization and the dipole aggregates, having a much smaller external electrical field than the free water molecules, preferentially associate with the apolar solute. This phenomenon is most commonly described by the statement that water molecules organize themselves around the apolar structures with a decrease in entropy with respect to bulk water. It would seem more appropriate to describe this phenomenon as one of selective interaction of the apolar solute with the self-neutralized molecular aggregates already present in the liquid. The electrically more neutral water aggregates increase with decreasing temperature and thus are responsible for the increase in solubility with decrease in temperature that has often been observed with apolar solutes.

Partition of Globular Proteins between Water and Another Phase

The simplest experiment that can be carried out to observe the transfer of a globular protein from water to an immiscible phase involves considerable limitations, as globular proteins do not show, as a general rule, appreciable solubilities in organic solvents and, as already discussed, the addition of organic solvents at low temperature is a classical method of protein precipitation and purification. The high solubility of globular proteins in water implies a corresponding insolubility, not only in those media almost completely immiscible in water, such as octanol or cyclohexane, but also in media such as water-saturated secondary butanol or methylethyl ketone, which contain approximately 35% of water. However, by a suitable modification of pH and the addition of appropriate ligands it becomes possible to transfer typical globular proteins to such relatively hydrophobic media as water-saturated butanol and pentanol. By lowering the pH to 2.5 the dissociation of glutamic and aspartic acids is repressed. It is well known that many globular proteins, lysozyme and serum albumin among them, do not undergo irreversible chemical reactions even after a long time at these acidic conditions. The disappearance of the negative charges results in a protein that is virtually a cationic polyelectrolyte. The molecular dipoles of water associate with the polyelectrolyte charges and these polar interactions are much stronger than those with the molecules of the more apolar media; as a result partition of the polyelectrolyte protein favors overwhelmingly the water phase. However, subsequent addition of an anion with some solubility in the "hydrophobic" phase progressively increases the solubility of the protein in this phase by replacing the interactions of the protein polyion with the water dipoles by ion pairs in which the positive charges of the protein become neutralized. As more such ion pairs are formed the protein–anion complex with its complement of hydrophobic amino acids acquires a partition coefficient that favors transfer to the apolar phase. The thermodynamic theory of the transfer of a protein between two phases under the influence of an added ligand can be done without explicit reference to the existence of charges or to the structural changes of the protein that may be brought about by the addition of ligands, including protons. The explanation of the observed effects in mechanistic terms may thus be reduced, as in other instances, to the interpretation of the energetic parameters extracted from a model-free description of the transfer.

Partition of Polyvalent Complexes between two Immiscible Phases

A protein, or in general any polymer, having N potential binding sites for a ligand X, is assumed to be able to partition between two phases, "upper" and "lower" [4]. The concentration of a particular protein–ligand complex PX_J in the lower phase is given by the Adair formulation (Chapter IV) as

$$[PX_J]_l = \binom{N}{J}[P]_l[X]_l^J \exp\left(\frac{\Delta G_l(J)}{RT}\right) \tag{4}$$

The concentration of PX_J in the lower phase, $[PX_J]_l$ is expressed in Eq. (1) as a function of the concentration of free protein and free ligand in the lower phase, respectively, $[P]_l$ and $[X]_l$, and the standard free energy of formation of the complex, $\Delta G_l(J)$, from 1 mol of protein and J mol of ligand, dissolved in the lower phase medium. An equation similar to (1) holds true for the concentration of PX_J in the upper phase,

$$[PX_J]_u = \binom{N}{J}[P]_u[X]_u^J \exp\left(\frac{\Delta G_u(J)}{RT}\right) \tag{5}$$

The overall partition coefficient β, between upper and lower phases, regardless of the distribution of the ligand among the protein molecules is

$$\beta = \frac{\sum_{J=0}^{N} \binom{N}{J}[P]_u[X]_u^J \exp[\Delta G_u(J)/RT]}{\sum_{J=0}^{N} \binom{N}{J}[P]_l[X]_l^J \exp[\Delta G_l(J)/RT]} \tag{6}$$

Let the standard free energy of transport of 1 mol of ligand *from lower to upper phase* be $\Delta G(X)$. The partition coefficient of the ligand is then

$$\alpha_X = \frac{[X]_u}{[X]_l} = \exp\left[\frac{-\Delta G(X)}{RT}\right] \tag{7}$$

From the definition of the direction of transfer of the ligand that we have adopted (lower phase $\rightarrow$ upper phase) it follows that if $\Delta G(X) > 0$ then $\alpha < 1$ and if $\Delta G(X) < 0$ then $\alpha > 1$. A similar rule applies to the free energy of transfer of the unliganded protein, $\Delta G(P)$ and the corresponding partition coefficient α_P:

$$\alpha_P = \frac{[P]_u}{[P]_l} = \exp\left[\frac{-\Delta G(P)}{RT}\right] \tag{8}$$

We define $\delta G_s(J)$ as the difference in the standard free energies of formation of the PX_J complexes in the two phases:

$$\Delta G_u(J) - \Delta G_l(J) = \delta G_s(J) \tag{9}$$

With the equivalences defined by Eqs. (7) to (9) the overall partition coefficient of the protein becomes

$$\beta = \frac{\alpha_P \sum_{J=0}^{N} F(X,J) \exp\{[-J\Delta G(X) + \delta G_s(J)]/RT\}}{\sum_{J=0}^{N} F(X,J)} \tag{10}$$

where we have set

$$F(X,J) = \binom{N}{J}[X]_l^J \exp\left[\frac{\Delta G_l(J)}{RT}\right] \tag{11}$$

A particularly simple form of the last equation is obtained if there is a unique free energy of stabilization for the binding of X in the upper phase, independent of J so that

$$\Delta G_s(J) = J\,\delta G_s \tag{12}$$

Setting then

$$\gamma = \exp\left[\frac{\delta G_s - \Delta G(X)}{RT}\right] \tag{13}$$

$$\beta = \alpha_P \frac{\sum\limits_{J=0}^{N} F(X,J)\gamma^J}{\sum\limits_{J=0}^{N} F(X,J)} \tag{14}$$

We note the very simple form of Eq. (14): Apart from the parameters that determine the binding of the ligand X by the protein in the lower phase, to describe the partition three additional parameters are required: the partition coefficient of the unliganded protein, the partition coefficient of the free ligand, and the difference between the standard free energies of binding of a mol of X in the upper and lower phases. If all N sites are equivalent and independent there is a single dissociation constant for the complexes in the lower phase, K and setting $z = [X]_l/K$ we then have by Eq. (IV.22)

$$\sum_{J=0}^{M} F(X,J) = (1+z)^N \tag{15}$$

$$\sum_{J=0}^{N} F(X,J)\gamma^J = (1+z\gamma)^N \tag{16}$$

and

$$\beta = \alpha_P \left[\frac{(1+z\gamma)}{(1+z)}\right]^N \tag{17}$$

Of the above mentioned three parameters that determine β in Eq. (17) α_P is usually inaccessible to experiment due to its very small magnitude, to the point that it is impossible to demonstrate any partition of the protein in the absence of the ligand. Consider, however, the Hill coefficient of the partition that we define as

$$H = \frac{d \log \beta}{d \log z} = \frac{z}{\beta} \frac{d\beta}{dz} = \frac{Nz(\gamma - 1)}{(1 + z)(1 + z\gamma)} \tag{18}$$

This is independent of α_P. We expect H to reach a maximum (H_{max}) in the region in which $z = 1$, and also to remain approximately constant over a large range of z. To determine the condition of maximum H we set $dH/dz = 0$, obtaining

$$\frac{dH}{dz} = \frac{N(\gamma - 1)(1 - z^2\gamma)}{[(1 + z)(1 + z\gamma)]^2} \tag{19}$$

$dH/dz = 0$ if $\gamma = 1$ or $z^2\gamma = 1$. The former alternative is excluded as it gives $H = 0$, so that the condition of maximum is

$$z = \frac{1}{\sqrt{\gamma}} \tag{20}$$

which introduced in Eq. (18) gives

$$H_{max} = \frac{N(\sqrt{\gamma} - 1)}{\sqrt{\gamma} + 1} \tag{21}$$

and

$$\gamma = \left[\frac{(N + H_{max})}{(N - H_{max})} \right]^2 \tag{22}$$

Since the partition coefficient of the ligand may be obtained by an independent experiment and

$$\gamma = \alpha_X \exp\left(\frac{\delta G_S}{RT} \right) \tag{23}$$

Eqs. (22) and (23) permit the determination of δG_S, the excess free energy of stabilization of ligand binding in the upper phase (organic solvent saturated with water) above that of the lower phase (water saturated with organic solvent). In experiments of Mustacich and Weber [4] lysozyme and serum albumin were partitioned at pH 2.4 between butanol and water using sodium *p*-toluene sulfonate as the ligand that neutralizes

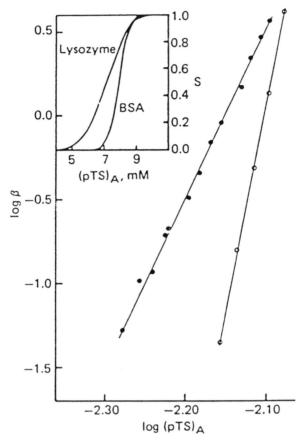

Figure 1. Hill coefficient in the partition of bovine serum albumin between water and butanol.

the cationic charge of the proteins. The partition coefficient of this ligand was 0.44. For bovine serum albumin the overall partition coefficient β was unity at a concentration of 7.9 mM p-toluene sulfonate. Direct determination of the ligand transported to the upper phase at virtually complete protein transfer was 80 mol of p-toluene solfonate per mol of serum albumin. The Hill coefficient measured (Fig. 1) was 24. From these data we obtain $\gamma = 3.64$ and $\delta G_S = -1.23$ kcal/mol. We note that when these values are introduced in Eq. **(17)** we obtain log $\alpha_P = 29.4$!. We also note the extremely narrow range of ligand concentration over which the transfer takes place. To anyone unfamiliar with the theory developed above this phenomenon may appear as one of extreme cooperativity requiring for its explanation perhaps some abrupt change in protein conformation that changes the partition coefficient of the protein when a critical number of ligands is bound. The theory shows that such supposition is unnecessary: The ligands are bound independently by the protein and each contributes the same increase $(-1.23$ kcal/mol$)$ to the free energy of transfer from the lower to the upper phase. The apparent critical effect arises simply from the fact that this increase in the free energy of transfer amounts to a change in the partition coefficient by a factor of 8, so that binding of one additional ligand is sufficient to change completely the solvent preference of a protein–ligand complex with a partition coefficient not much smaller than 0.1.

An aspect of considerable interest is that the stability of ion pairs and therefore their ability to induce transfer to an apolar phase increases with the difference in pK of the neutralizing groups. Thus sulfonic groups can be expected to superior in this respect to carboxyl and phosphate. The presence of sulfonic groups in some pharmacologically active peptides such as gastrone or fibrinopeptide may perhaps be related to this advantage. In the same vein the ion pairs formed by a given anion with arginine will be more stable than those formed by lysine. There is ample evidence that the preference of arginine over lysine in many circumstances of biological interest arises from this property [5, 6].

Under the conditions described above, pH 2.5 and addition of p-toluene sulfonate, transfer of proteins into butanol could be achieved though transfer to pentanol or hexanol did not prove possible. It was reasonable to expect partition in a less polar solvent at low pH, or in 1-butanol at a higher pH by increasing the number of nonpolar groups in the protein through chemical modification, or by altering in similar fashion the pK of the anion-binding groups and thereby stabilizing ion pairing, or by employing a ligand with a lower acidic pK or a partition coefficient more favorable to the apolar medium (e.g., octylsufonate). To some extent these various possibilities have been examined [7]: On partial ethylation of the carboxyl groups with triethyloxonium fluoroborate a preparation of albumin was obtained that could be transferred to butanol at a concentration of p-toluene sulfonate that was one-half of that required for the transfer of the intact protein, but the order of the transfer reaction was ≈ 7 instead of 25. To obtain ion pairs of greater stability the ϵ-amino groups of lysine were changed to homoarginines by treatment with O-methyl urea. In these modified proteins (ethylated and/or guanidated) a fraction of the protein separated at the interface, a behavior that must be attributed to the heterogeneous protein population that the chemical treatment originated. Though not further characterized these protein fractions that remain at the interface at the partition experiments present considerable interest: They can undergo partial transfer into the nonpolar phase because some parts of the protein interact strongly with the apolar phase and others with water, neither interaction being sufficiently strong to overcome the other. They thus offer us a prototype of the integral membrane protein. One important question to be answered is that of the physical state of the proteins transferred to the polar solvent; ultracentrifugation observations in phases of mixed solvents are difficult to interpret, but some information can be obtained by a study of the rotational diffusion of dansylated conjugates. In the experiments described [7] the polarization of the emitted fluorescence was found to be independent of concentration indicating the absence of aggregation in the apolar medium. This is not surprising because of the considerable residual charge of these proteins and the low dielectric constant of the solvent. The observations of the polarization of the fluorescence can be given a straightforward interpretation because the linear dependence of the reciprocal of the polarization on the ratio temperature/viscosity was followed in all cases over a range of temperatures of 2 to 38°C [4, 7]. It is uniformly found that the rotational relaxation times computed from the Perrin plots correspond to molar volumes that are one-half to one-quarter of that in water. This decrease in apparent volume may be rationalized as arising from the association of the alcohols with the nonpolar residues of the protein. This association breaks internal protein bonds and facilitates the independent rotations of some regions of the molecule with respect to others. It has been observed that the S100 protein can be transferred to pentanol at *neutral pH* by the addition of salts of divalent (Ca^{2+}) and even monovalent cations, and in these cases a notable reduction of the rotational relaxation time is also observed [8]. S100 is known to associate with lipid vesicles inducing in them as increased cation leakage [9] but it is noteworthy that this association is accompanied by an increase in rotational relaxation time indicating an interaction of the protein with the surface of the vesicles rather than penetration into the lipid phase [8]. One would conclude from these observations that the independent motions of parts of the protein are increased to a variable extent when a protein interacts with a membrane and that the more extensive the contacts with the lipid phase the larger the independent motions. Although the protein may show a compact well-defined structure in a crystal or in water solution, association with a membrane may be expected to increase the underlying polymorphism. It is a general difficulty encountered in all branches of chemistry that the molecular states directly involved in the chemical reactions are not accessible to direct observation on account of their rarity in comparison with the inert "ground" state. A very small energy gap (2–3 kcal) between these states is all that is necessary to make such active species undetectable. We may find that in the intrinsic membrane proteins uncertainties of this origin are compounded by the polymorphism acquired by the protein through its intimate contact with the lipid.

The Partial Transfer of Proteins across Membranes under the Influence of Ligands

The studies of the integral membrane protein carried out in the past 15 years have shown that they share a common overall structure: Parts of the protein are submerged in the membrane making direct contact with the lipid, whereas other parts, comprising the two ends of the peptide chain, remain in the two aqueous phases separated by the lipid bilayer. Attached carbohydrate moieties often serve the purpose of limiting further penetration of the protein into—or escape through—the membrane by their strong hydrophilic character [10]. This structure is remarkably well suited to permit changes in the interaction of the protein and membrane under the influence of ligands: The partition of significant portions of the peptide chain between the lipid bilayer and the aqueous phase may be expected to depend on the binding of specific ligands providing in this way for the influence of ligands of the external medium on the functions of the membrane proteins in the cell. The observations of the overall transfer of globular proteins from an aqueous to a hydrophobic phase under the influence of ligands described above provide some indication of the structural origin and the energetics of such phenomena. To begin with, we expect that the larger changes in the partition coefficient will be those brought about by the neutralization of charges through the formation of ion pairs. The larger the difference in pK between the acidic and basic groups that form the neutral pair the larger the free energy of neutralization and the consequent stabilization on transfer to a nonaqueous phase. In this respect we expect the arginine residues of the protein to furnish most often the basic moiety. The protein-transfer studies show that the gain in the free energy of transfer resulting from the formation of a single ion pair of a basic amino acid with a p-toluene sulfonate residue amounts to somewhat more than a kilocalorie per mole. Therefore, its formation is sufficient to promote appreciable to virtually complete transfer of a complex to the hydrophobic phase if the system is poised, that is if the partition previous to binding favors the water phase by a factor smaller than 10. We find again in this case that the free energy changes required for the switching on, or off, of the physiological functions through ligand binding are very small, of the order of $2RT$ or $3RT$ and therefore that the calculation of the effects, by application of the principles of molecular interaction to the structural data, is rendered very uncertain.

The investigations of integral membrane proteins have been to now concerned with the structural aspects, with very little emphasis on the energetics and dynamics, although the sensitivity of transfer to the influence of ligands, of which examples have been discussed above, indicates that phase transfer must be of the greatest functional importance in this area of research. It has been noticed [11] that bacteriorhodopsin shows a set of helical serial regions that traverse, side by side, the purple membrane. Following this observation a similar structural possibility has been deduced for a number of intrinsic membrane proteins of which the sequence is known [10]. Engelman *et al.* [12] have calculated that the free energy of formation of a helical segment in a nonpolar medium is much lower than that in water, thus providing the necessary stability for the transmembrane protein segments. This has given an important starting point in the evaluation of the structural disposition of proteins that are integral to the membrane, but virtually nothing is known of the dynamics associated with their function. Progress in this area will require the study of thoroughly well-defined model systems of increasing complexity, an area that so far has attracted little interest. However, without excessive speculation we may foresee how phase partition changes may be responsible for one of the most important observations of cell physiology [13]: the transmission of physical influence across membranes through the integral membrane proteins.

Physical Communication across Cell Membranes

A simple, perhaps the simplest, way in which the effects of ligands bound external to the cell membrane may affect the chemical reactions occurring in the interior of the cell is schematically depicted in Figure 2. A protein that spans the cell membrane has compact sequential portions of the peptide chain I, M, and O. M is an integral protein portion that can bind either I or O, but not both simultaneously. The fractional concentrations of I and I_i inside the cell, I_m within the membrane, and IM as a complex with M. Similarly the fractional concentrations of O are O_x

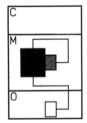 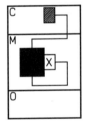

Figure 2. Transfer of portions of protein from the medium to the cytosol across a membrane under the influence of ligands. Binding of X displaces a protein segment which is then transferred to the cytosol.

outside the cell, O_m inside the membrane, and OM in complex with M. I and O can effectively partition between the membrane and the aqueous media, inside or outside the cell respectively, with corresponding partition coefficients k_i and k_x:

$$\frac{I_m}{I_i} = k_i; \qquad \frac{O_m}{O_x} = k_x \qquad (24)$$

The equilibrium of M with the peptide chain segments O and *I* depends on the equilibrium constant K_D of the disproportionation reaction $I_m + OM \leftrightarrow IM + O_m$:

$$K_D = \frac{I_m OM}{O_m IM} \qquad (25)$$

We expect that the forms M, I_m, and O_m are negligible fractions of I and O, respectively, so that to a sufficient approximation:

$$IM + I_i = 1; \qquad O_x + OM = 1; \qquad IM + OM = 1 \qquad (26)$$

From Eqs. (25) and (26) we then have

$$\frac{MI}{(1 - MI)} \frac{(1 - OM)}{OM} = \left[\frac{MI}{(1 - MI)} \right]^2 = \frac{K_D k_i}{k_x} \qquad (27)$$

and

$$\left[\frac{MI}{(1 - MI)} \right]^2 = \left[\left(\frac{1}{I_i} \right) - 1 \right]^2 \qquad (28)$$

which yield

$$I_i = \left[1 + \left(\frac{K_D k_i}{k_x} \right)^{1/2} \right]^{-1} \qquad (29)$$

From Eqs. (29) and (27) we may derive the fractions of the system in the two alternative states shown in Figure 2. We may expect that certain physiological functions inside the cell depend on the fraction I_i, which can, for example, have the property of activating a specific catalyst. We expect K_D to be in the vicinity of 1, as this will confer maximum sensitivity to the system, as regards the effect of changes in k_i and k_x. If $k_i/k_x = 1/100$, 91% of I is bound to M within the membrane and 9% appears in the inside of the cell, while the percentages for O are complementary to these. If the partition coefficient for O is increased by the binding of a specific ligand and the ratio k_i/k_o is inverted then 91% of I will now be found within the cell.

The Effect of Pressure on the Transfer of Ligands between Phases and on the Membrane Functions

The solubility of a substance in a solvent can be affected in various ways by an increase in the hydrostatic pressure applied to the system. Increase or decrease of solubility with pressure will depend on the relative effects on solvent–solvent, solute–solute, and solute–solvent interactions. If a solute is present in two phases in contact, at a concentration that is well below saturation at all the applied pressures, the effect of pressure on the solute–solute interactions may be disregarded. The effects will therefore be determined, in the main, by the pressure dependence of solvent–solvent and solvent–solute interactions. A solute that is already present at atmospheric pressure at measurable concentrations in two media of the composition of the membrane and the neighbor aqueous solution is one that is held, of necessity, by van der Waals forces in the two of them. Both solvent–solvent and solute–solvent interactions will be increased by the shortening of the intermolecular distances because of compression but packing factors will make the increase in solvent–solvent interactions the dominant influence: The pressure will act to "squeeze out" or reject the solute. This effect will increase with the compressibility of the solvent and as a result the relatively larger rejection from the hydrophobic phase will drive the solute toward the hydrophylic phase in contact with it. Most probably this is the mechanism by which the effect of various anesthetics is relieved by hydrostatic pressure [14].

The paucity of the experimental results with simple model systems prevents us from making even a crude estimate of the magnitude of the effects that are to be expected on application of pressures in these cases. More interesting for our purpose is the consideration of the effect of pressure on the transfer of molecular complexes. Here the situation is both more involved and more interesting. Besides the van der Waals forces, molecular complexes can be held together by electrostatic interactions of permanent partial charges, which include hydrogen bonds, and by formation of ion pairs. Let A and D stand for the partners entering into the complex AD. In the polar solvent the equilibrium between free partners and complex involves the presence of charges from either partners or solvent and may be formulated as

$$A \mp S + D \pm S \leftarrow K_S \rightarrow ADS \pm \qquad (30)$$

where ADS $\pm$ indicates the involvement of charges in the complex and solvent in some fashion. In the apolar solvent the possible charges of the partners will be neutralized in the complex and the equilibrium is of the form

$$A \mp S_0 + D \pm S_0 \leftarrow K_A \rightarrow ADS_0 \qquad (31)$$

Evidently the complex must have a smaller dissociation constant in the apolar solvent ($K_A < K_S$). The results that follow the application of pressure depend on the relative changes of K_A and K_S with pressure and therefore on the volume change upon formation of the complex. In the polar solvent we expect that the separate partners will occupy a smaller volume than the complex because of the shorter bonds involved in the electrostatic interactions. This effect will be particularly marked when A and B carry charges of opposite sign so that AB is an uncharged ion pair that is separated by the solvation of the charges with consequent electrostriction of the solvent. In these cases application of pressure will favor the transfer of complex from the apolar to the polar medium. We note then that in the case of both apolar compounds and complexes involving charges pressure will favor the partition to polar solvent. In this respect it is of interest to note the much larger sensitivity to pressure of complex physiological functions that involve membranes, in comparison with those effects that are observed in macromolecules in water solution (Chapters XII and XV).

There are two different factors responsible for this larger pressure sensitivity: The internal pressure of water is nearly 10-fold greater than that of a typical aliphatic hydrocarbon such as hexane and it is to be expected that the effects of the added pressure will be larger the smaller the internal pressure of the liquid. A second important difference is that the perturbation of a physiological function arises from effects on a single, dominant rate of an ongoing, usually complex kinetic process, whereas on application of pressure to a macromolecule in equilibrium with its surroundings the perturbation must necessarily affect the opposing rates that determine the thermodynamic equilibrium. Thus, in distinction with the much higher pressures necessary to perturb the structure of proteins, we find that reversal of local or general anesthesia takes place with pressures of about 150 atmospheres [14], and that opening and closing of ion channels [15] and conspicuous changes in transmission of the nerve impulses [16] require pressures of only 200 to 300 atmospheres. In the same range we find the inhibition of catalysis by the membrane ATPases by hydrostatic pressure [17]. We may advance the hypothesis that the cases most sensitive to pressure primarily depend on the changes in partition of molecules, or portion of molecules, between membrane and surrounding water or, more explicitly, that they follow partial or total rejection of the molecules from the membrane phase when pressure is applied. A confirmation of this supposition must await experiments with model systems including the transfer of peptides and small proteins between well-defined phases. Indeed we cannot expect progress in this area before the effects that we have crudely analyzed above are subject to detailed examination using the simplest possible systems.

References

1. Hildebrand, J.H., and Scott, R.L. (1950) *The Solubility of Non-electrolytes*. Reinhold, New York, pp. 94–105.

2. Rahman, A., and Stillinger, F.H. (1971) *J. Chem. Phys.* **55**, 3336–3359. Rahman, A., and Stillinger, F.H. (1973) *J. Am. Chem. Soc.* **95**, 7943–7948.

3. Review of studies of water by molecular dynamics: Stillinger, F.H. (1980) *Science* **209**, 451–457.

4. Mustacich, R.V., and Weber, G. (1978) *Proc. Natl. Acad. Sci. U.S.A.* **75**, 779–783.

5. Riordan, J.F., McElvany, K.D., and Borders, C.L., Jr. (1976) *Science* **195**, 884–886.

6. Padlan, E.A., Davis, D., Rudikoft, S., and Potter, E.M. (1976) *Immunochemistry* **13**, 945–949.

7. Mustacich, R.V., and Weber, G. (1979) *Biochemistry* **19**, 980–985.

8. Morero, R.D., and Weber, G. (1980) *Biochim. Biophys. Acta* **703**, 241–246.

9. Calisano, P., Alema, S., and Fasella, P. (1974) *Biochemistry* **13**, 4553–4560.

10. Gennis, R.B. (1989) *Biomembranes*, Chapter 3. Springer-Verlag, New York.

11. Henderson, R., and Unwin, P.N.T (1975) *Nature (London)* **257**, 28–32.

12. Engelman, D.M, Steitz, T., and Goldman, A. (1986) *Annu. Rev. Biophys. Biophys. Chem.* **15**, 321–353.

13. Levitzki, A. (1984) *Receptors*. Part 3: Formal Kinetics of Receptor Systems. Benjamin/Cummings, Menlo Park, CA.

14. Lever, M.J., Miller, K.W., Paton, D.M., and Smith, E.B. (1974) *Nature (London)* **231**, 368–371. The original observation of the inhibition of the effects of anesthetics by pressure was in luminescent bacteria: Johnson, F.H., Brown, D., and Marsland, D. (1942) *Science* **95**, 200–203.

15. Conti, F., Heinemann, S.H., and Stuehmer, W. (1987) In *Current Perspectives in High-Pressure Biology*, H.W. Jannasch, R.E. Marquis, and A.M. Zimmerman, eds. Academic Press, London, pp. 171–179.

16. Spyropolous, C. (1957) *J. Gen. Physiol.* **40**, 848–857 made the important additional observation that alcohol narcosis of the nerve was relieved by a pressure of 500 atmospheres or reduction of the temperature by 15°C.

17. Chong, P.L.-G., Fortes, G., and Jameson, D.M. (1985) *J. Biol. Chem.* **260**, 14484–14490. Heremans, K., and Wuytack, F. (1980) *FEBS Lett.* **117**, 171–173.

XII

Detection and Measurement of the Statics and Dynamics of Protein–Ligand and Protein–Protein Associations

Interactions of Proteins and Small Ligands

Measurements of the degree of dissociation of protein–ligand complexes are done employing direct, semidirect, and indirect methods. Direct methods are those in which the concentration of bound ligand is determined by its contribution to some intrinsic property of the solution that depends on the fractions of free and bound species. The measurement of coligative properties and various spectroscopic techniques belong in this category. In the semidirect methods [1] a further equilibrium of the solution with a separate phase is created—by the use of a semipermeable membrane, an immiscible solvent, or a highly cross-linked polymer—in which one of the interacting molecular species, usually the protein, has effectively zero concentration. In the indirect methods use is made of kinetics, in one form or another. The most successful application of the indirect methods involves the separate determination of the rates of association and dissociation [2]; in yet other cases the rate of an unrelated reaction known to depend on the concentration of either free or bound ligand is measured [3]. Whereas direct or semidirect methods depend only on the thermodynamic parameters of the system the indirect methods always involve assumptions regarding the relations of these to the kinetic processes that are measured. The direct methods are the most accurate, though not always the simplest or most reliable, but under the best conditions the semidirect methods are fully comparable.

In applying any method one must remember that we are always measuring the average value of a macroscopic property of the system resulting from an unknown distribution of that property over the molecular population. Two different methods of measurement that do not weigh equally the members of the population will report different values for the derived thermodynamic parameters, and it may be surmised that this will always be the case to some extent. The average value for the free energy of association will be independent of the instantaneous fluctuations, but the value derived from a measured parameter (e.g., spectroscopic) will be influenced by the effects produced by the fluctuations and will faithfully represent the energy average only if it is rigorously linear with the energy over the range of values covered by the energy distribution. Differences in the measured average are likely to be particularly important in the case of the weaker complexes, because the reduced forces between the partners predicate a variety of possible interactions. The thermal energy will disrupt the complexes to the extent of changing the energy of interaction of the molecules by $\pm kT$ and as the free energy of association becomes smaller this amount will represent a larger fraction of the total.

In complexes of proteins with small ligands the situation is even more complicated: The protein site that binds a small ligand is maintained by a number of weak interactions that are themselves subjected to thermal fluctuations, and this circumstance will result in a larger heterogeneity of binding. Additionally the small bound ligand may adopt a conformation that is not the one obtained when free in solution as exemplified by a recent NMR study of acetylcholine receptor complexes [4]. Even in the cases in which the total free energy of binding is very large, as in the heme bound to apohemoglobin, the multiple interactions with the protein may not result in a unique disposition of the ligand as shown by the NMR observations of La Mar and co-workers [5].

No single method of investigation can be applied to determine the whole range of free energies of association exhibited by biological molecules. The smaller values, of 2 or 3 kcal mol^{-1}, are observed in the binding of substrates to enzymes and in the unspecific binding of ions to proteins. There is considerable difficulty in their investigation because the ligand concentration required to obtain some association would usually be too large in comparison with that of the protein to permit the use of semidirect methods such as equilibrium dialysis, and the energy of interaction with the protein may be too small to give rise to spectrally detectable effects. Most of the values that are attributed to these weaker complexes are derived from activation or inhibition of catalytic properties of the proteins rather than by direct measurements of the degree of dissociation. Spectroscopic methods are particularly useful when the ligand is an aromatic compound that binds to the protein with dissociation constants that may vary from micromolar to picomolar. In the associations that involve free energy changes of 16 kcal or more all the known values have been determined by indirect methods [2, 3].

Methodologies for the Study of Protein–Protein Equilibria

To determine the free energy of noncovalent association of protein subunits into aggregates it is necessary to measure the degree of dissociation at concentrations of the order of the dissociation constant of the aggregate. The classical method of determination of the state of aggregation of dissolved substances is through the coligative properties of the solutions, but these involve concentrations that exceed by several orders of magnitude those in which there is appreciable dissociation of the protein aggregates. Additionally at such concentrations no solution of macromolecules would have ideal thermodynamic behavior. In spite of these difficulties, measurements of osmotic pressure were instrumental in the initial determinations of the mass and the state of aggregation of proteins [6, 7]. The dissociation of many oligomeric proteins was originally demonstrated by the study of their sedimentation equilibria in the ultracentrifuge [8]. However, in media of physiological ionic strength most protein aggregates do not show appreciable dissociation when observed at the lowest concentrations ($> 10^{-6}$ M) at which the sedimentation properties can be reliably determined in the ultracentrifuge. Indirect methods of various kinds, including enzyme activity measurements [9], have been used to demonstrate that the concentration range at which dissociation of many aggregates becomes detectable cannot be greater than 10^{-7} M and must be in many cases close to or smaller than picomolar.

Importance of Spectroscopic Methods

Because of the shortcomings of the methods that directly measure the mass or number of the particles, detection of protein dissociation is restricted to indirect methods that permit the observation of concentration-dependent changes in various macromolecular properties. In principle the concentration dependence of any dissociation-sensitive spectroscopic property can be used to determine the degree of dissociation and therefore the free energy of association of the oligomers in a protein. However, many spectroscopic procedures that could conceivably be useful to detect these changes such as NMR and ESR, Raman scattering, or infrared absorption are of exceptional application as they are currently limited to observations at concentrations much higher than the dissociation range of most aggregates. The scattering, absorption, or emission of ultraviolet or visible light, and conceivably the procedures that detect nuclear disintegration [10], are the only methods of general application that can reach the required low concentrations. The change in spectroscopic properties on change in state of aggregation resulting from intrinsic differences of aggregate and subunits may be accompanied by additional changes that are independent of it. These arise from the perturbation that brings about the dissociation rather than from the dissociation itself. Adsorption of a significant fraction of the protein to the walls of the containing vessel by changes in temperature, pressure, ligand addition, or dilution is among such sources of change in the spectroscopic properties. Therefore to attribute a spectroscopic change to dissociation it is indispensable to ascertain that it varies with the concentration of the protein in a manner that can be reasonably attributed to the existence of an equilibrium of second or higher order of a dissociating system. We note that at the high dilutions necessary to achieve a measurable dissociation equilibrium in oligomeric proteins the average near-neighbor distances between particles [11] are hundreds to thousands of Å units, so that long-range concentration-dependent particle interactions may be excluded with certainty.

To calculate an intermediate degree of dissociation α between complete dissociation ($\alpha \to 1$) and complete association ($\alpha \to 0$) one needs to determine the limiting values of the spectroscopic property at high and low dissociations, respectively, φ_1 and φ_0. In practice these are obtained from "plateau" regions, that is, from the regions of nearly constant value of the spectral property at the beginning and end of a plot of the spectral value against the variable that determines the dissociation: protein or ligand concentration, temperature, or pressure. If $\varphi(\alpha)$ denotes the value of the spectral property at an intermediate degree of dissociation

$$\varphi(\alpha) = (1 - \alpha)\varphi_0 + \alpha\varphi_1 \tag{1}$$

and consequently

$$\alpha = \frac{\varphi(\alpha) - \varphi_0}{\varphi_1 - \varphi_0} \tag{2}$$

According to the above equations the φ property is linearly related to the proportions of the components. The limitation thus imposed may be illustrated by comparing the determination of the dissociation by measurements of light absorption with those of fluorescence yield. The intensity of the fluorescence emission depends not only on the fractional absorption of light at the chosen excitation wavelength, but also on the relative yields of the fluorescence emission of the aggregate and subunits, q_0 and q_1, respectively. Equation (1) then becomes

$$q(\alpha)\varphi(\alpha) = q_0\varphi_0(1 - \alpha) + q_1\varphi_1\alpha \tag{3}$$

Evidently knowledge of the apparent fluorescent yield alone at an intermediate degree of dissociation is insufficient to determine α. Nevertheless if excitation is carried out at an isosbestic point of the absorption spectra of aggregate and subunits, then $\varphi_0 = \varphi_1$ and the fluorescence yield becomes linearly dependent on the degree of dissociation. Equations (1) and (2) are now valid when φ_0 and φ_1 are replaced by q_0 and q_1, respectively. Equations (1) and (3) presuppose that at intermediate degrees of dissociation the aggregate and the dissociated species then present are still characterized by the values φ_0 and φ_1 assigned to them from observations of the solutions in which one of those species is by far the predominant form. As we indicate later there is very good evidence that this is not generally the case and that Eq. (2) should be used as a reasonable approximation, to be evaluated in each circumstance, rather than as an exact formulation.

Absorption and Fluorescence Spectra

Changes in the absorption and fluorescence spectra on dissociation of a protein aggregate may be expected on two different grounds: first, following dissociation the environment of the amino acid residues at the boundary between subunits will be partially replaced by solvent; second, we expect that separation of the subunits will be followed by conformational rearrangements that change the immediate environment of the fluorophore even if this is not in direct contact with solvent. In water solutions the contacts of the subunit boundary groups exposed on dissociation are with strongly polar water molecules, an environment that differs considerably from the apolar immediate vicinity that predominates in the aggregate. The effects of the polarity of the medium on the absorption and fluorescence spectra of aromatic organic compounds have been much studied. The classical theory of the effects of solvent polarity on the absorption of light [12] treats the problem as one of average energy conservation: The changes in the electronic energy levels derive entirely from compensating changes in the energies of molecular interaction. The relations between changes in the electronic levels and the interaction with solvent are shown in Figure 1: The electronic energy levels in the gas phase (A) are modified in the solvent (B, C) by interactions that depend on the nature of the electronic distribution in the chromophore in both ground and excited states, as well as the character of the solvent molecules. One must consider the effects resulting from both permanent and induced charges. The latter will exist in all cases because of the universal character of the London–van der Waals forces. If the ground and excited state were stabilized by these forces to an equal extent the absorption and emission

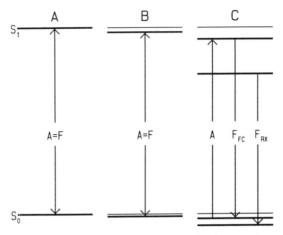

Figure 1. S_0, singlet ground state; S_1, lowest singlet excited state of a molecule. Transitions between these states in the gas phase (A), apolar solvent (B), and polar solvent (C). The arrows join the lowest vibrational state of each level. Notice that the solvent interactions of a polar molecule stabilize S_0 as well as S_1.

spectra in gas phase and solvent should be identical. Observation indicates that this possibility is not realized in practice; In an apolar solvent, such as cyclohexane, the spectra are generally displaced to the red some 100 cm^{-1}, equivalent to $kT/2$ at room temperature, with respect to the gas spectrum [13]. This "universal red shift" indicates that the molecules in the excited state have a considerably larger polarizability than in the ground state, as would be expected from general principles of molecular physics. Chromophores in solution give rise to very broad, usually unresolved absorption and emission bands, so that in applying a principle of energy conservation to them the total energy of the electronic absorption or emission must be integrated. Application of this rule in the case of fluorescence is straightforward as the fluorescence emission consists, in the overwhelming majority of cases, of a single, continuous band that results from transitions to the ground state from a unique electronic state. It is customary to define a center of mass of the emission, $\langle \nu_G \rangle$ by means of the equation:

$$\langle \nu_G \rangle = \frac{\sum F_i \nu_i}{\sum F_i} \tag{4}$$

where F_i is the photon intensity at wavenumber ν_i and the sums in Eq. (4)

are carried over the range of appreciable values of F. In contradistinction, the absorption spectra of chromophores in solution consists of a virtual continuum that starts in the visible or ultraviolet regions and continues toward shorter wavelengths. In this continuum there are peaks and valleys that permit only an artificial separation of the spectrum into bands that may or may not correspond to separate electronic transitions. This ambiguity makes it difficult to obtain reliable results by an integration such as that of Eq. (4), particularly when the observed spectral differences are small.

Figure 1B shows the *average values* of the electronic energy levels of ground and excited state in an ideal apolar solvent in which only the change in polarizabilities introduces differences with the gas-phase levels and Figure 1C those in a strongly polar medium such as methanol or water. The considerably greater displacements of the absorption and emission depicted for the latter case are observed only in "polar" fluorophores, that is, those with an electronic distribution in the excited state that differs greatly from the distribution in the ground state. The exact calculation of the effects expected for this case would require an estimation of the contribution of charge interactions from each atom in the chromophore molecule. However, the present description of the environmental effects on the spectroscopic properties must content itself with a limited analysis that considers the molecules as dipoles with different electric moments in ground and excited state, μ and μ^*, respectively [14]. In a polar solvent the orientation of the solvent dipoles with respect to the chromophore dipole of strength μ sets up previous to excitation a reaction field R_0, which derives from the orientation of the polar molecules of the solvent with respect to the dipole chromophore. Neglecting the universal dispersion forces, in an ideal apolar solvent $R_0 = 0$. The interaction of reaction field and dipole decreases the energy of the ground state by the amount μR_0. To compute the effects on the excited state it is necessary to recall that electronic transitions are governed by the Franck–Condon principle according to which the changes in electronic energy associated with the absorption or emission are virtually instantaneous in comparison with the much slower changes in nuclear coordinates. The sequence of steps that follow the absorption of the exciting light and continue with the emission of fluorescence is shown in Figure 2. The state reached in absorption has an energy loss $\mu^* R_0$ with respect to the apolar solvent. This results from the interaction of the new excited state dipole moment, μ^*, with the lagging reaction field, R_0, established previous to excitation. The absorption is shifted to the red by an energy $R_0(\mu - \mu^*)$ with respect to the absorption in an apolar solvent. Subsequent equilibration that further decreases the energy of the excited level to the value $\mu^* R^*$ depends on sufficient reorientation of the solvent dipoles during the limited duration of the fluorescent state. As R^* corresponds to a more

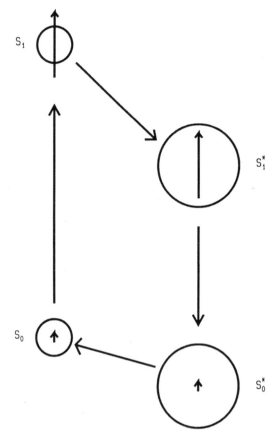

Figure 2. Interactions of a polar chromophore
with a polar solvent, in the ground state
and after absorption of light and fluorescence. S_0, ground state characterized by (R_0, μ); S_1, Franck–Condon
excited state reached in absorption
(R_0, μ^*); S_1^*, fluorescence relaxed state
(R^*, μ^*); S_0^*, Franck–Condon state following fluorescence (R^*, μ).

favorable orientation of solvent and chromophore dipoles, R^* can be
much larger than R_0. The excited state generated on absorption is called
the Franck–Condon excited state; the state following partial or complete
solvent reorientation is called the relaxed excited state. The solvent
reorientation that produces the change in reaction field from R_0 to R^* is

completely inhibited in a vitrified solvent and in this case R_0 persists and the shift of the fluorescence with respect to an apolar solvent can be identical to the absorption shift. In a fluid solvent in which all necessary reorientation of the solvent dipoles can be accomplished during the fluorescence lifetime the red shift of the emission reaches the value $R^*(\mu - \mu^*)$, and this may exceed the absorption shift by several thousand cm^{-1}. Thus not only are the measurements of emission shifts easier to quantitate than those of absorption, but their magnitude is considerably greater making them superior in the detection of changes of environment and conformation.

The qualitative effects resulting from changes in chromophore environment that follow on those in the state of protein aggregation are easy to interpret by analogy with the differences seen in solvents of different polarity. However, the quantitative computation in terms of interactions of the fluorophore with solvent or protein charges is much more difficult: Instead of the statistical distribution of charges of the pure solvent we have in the protein environment a precise topography of charge distribution with respect to the fluorophore. In a restricted number of cases this topography is known from X-ray analysis of the protein crystals, and the fixed character of the environment ought to permit, in the long run, an analysis of the spectroscopic properties far more detailed and penetrating than that possible today from observations of the chromophore in the less well-defined surroundings of liquid solvents. However, in the majority of cases the microscopic protein environment of the chromophores is not exactly known and calculation of R_0 is rendered very uncertain. Calculation of R^* would be even less precise as it involves possible motions of the protein charges during the excited state. These motions will be appreciable as the forces between the nearby dipoles and the monopoles of the excited fluorophore are likely to be several times larger than the thermal energy [14, 15]. The difficulties involved are exemplified by the observations of Macgregor and Weber [16] on the complex of the polar fluorophore 2-dimethylamino-6-cyclohexanoyl-4′-carboxylate (DANCA) with myoglobin. From observations of the fluorescence spectrum in diverse solvents the dipole moments of DANCA in ground and excited state are estimated to be 2 and 20 D units, respectively. In micromolar concentrations DANCA binds at the heme pocket of sperm-whale myoglobin as shown by its stoichiometric release on addition of hemin. In the protein the probe shows a shift in $\langle \nu_G \rangle$ of absorption of ~ -1700 cm^{-1} and a virtually equal fluorescence shift at $-74°C$. This corresponds to a medium of the polar characteristics of dimethylformamide or N-methylacetamide, solvents that one would expect to mimic the polar characteristics given by the peptide bonds to the protein interior. At room temperature the fluorescence spectrum shows an additional red shift of 1200 cm^{-1} indicating the conversion from R^0 to R^* as a result of relaxation of either solvent or protein charges. On the assumption that the probe occupies the plane of the heme and that it does not appreciably disturb the structure of the protein it was calculated that interaction of the polar chromophore with the amide dipoles of the protein could account for most or all of the low-temperature displacement of the spectrum.

Detection of Dissociation by Fluorescence Methods

Fluorescence methodology finds a natural application to the study of the reversible aggregation of proteins, as it can be readily applied to solutions of fluorophores at concentrations of 10^{-6} to 10^{-11} M. The concentration-dependent changes in spectral distribution, quantum yield, lifetime, or polarization of the fluorescence of protein solutions can be used to follow the dissociation. Polarization observations either from the intrinsic fluorophores of the protein, the tyrosine and tryptophan residues, or from covalently attached fluorescent labels have the added advantage that they depend directly on the rate of rotation of the fluorophore and thus contain information about the motional properties of the fluorescent unit.

Hierarchy of Motions in Protein Molecules

The rotational motions of the fluorophore depend on the bonds that link it to the rest of the particle. If these were strong enough to remain unchanged during the lifetime of the excited state, the fluorophore would undergo rotations determined exclusively by the frictional resistance offered by the surrounding medium to the motions of a rigid body of the dimensions of the protein particle. However, the bonds that link the various parts of the protein are, in the physiological range of temperatures, weak enough to permit independent motions of portions of the protein. Conceivably these local motions may involve a volume as small as the chromophore itself (residue motions) or a larger fraction of the particle (domain motions). The fluorescence properties can be modified only by those motions that are accomplished during the fluorescence lifetime, which typically lasts a few nanoseconds. In compact globular proteins only the overall motion of the particle and domain motions that involve the fluorophore alone or a slightly larger volume can reach appreciable amplitudes during these times. Domain motions of larger portions of the protein require breaking of many elementary bonds between the amino acid residues and are thus very unlikely to become significant during the few nanoseconds of the fluorescence lifetime.

The Effects of Local and Whole Particle Rotations on the Polarization of Fluorescence

Granted that the only motions appreciable during the fluorescence lifetime are the overall motions of the particle as a rigid—or quasirigid—body, and local rotations that involve the fluorophore alone we shall briefly characterize their effects on the polarization of the emitted fluorescence. In the absence of motion of any kind the linear polarization of the fluorescent light reaches its maximum value, p_0. Depolarization of the fluorescence results from the change in orientation of the emission oscillator of the fluorophore during the fluorescence lifetime. If the average orientations of the emission oscillator at the times of excitation and emission determine an angle Θ then the loss of polarization is given by the relation [17]

$$(1/p - 1/3) = (1/p_0 - 1/3)\frac{2}{3\cos^2\Theta - 1} \tag{5}$$

where p is the observed polarization. It is sometimes convenient to replace the polarization p by a related function, the anisotropy $a = (2/3)(1/p - 1/3)^{-1}$, in terms of which the last equation becomes

$$r = \frac{1/p - 1/3}{1p_0 - 1/3} = \frac{a_0}{a} = \frac{2}{3\cos^2\Theta - 1} \tag{5'}$$

With an accuracy sufficient for our purpose the overall particle motions are described by an equation first derived by Perrin in 1926 for the depolarization of the fluorescence by isotropic rotations of a rigid body. For this case [18] the depolarizing effect of the motion is given by

$$r = 1 + \frac{6RT\tau}{f_{\text{rot}}} \tag{6}$$

Equation (6) describes the polarization of the fluorescence emitted in a direction normal to both the direction of the electric vector of the polarized exciting light and its direction of propagation; τ is the fluorescence lifetime, R the gas constant, T the Kelvin temperature, and f_{rot} the frictional coefficient for rotation. For a rigid spherical particle

$$f_{\text{rot}} = 6\eta V \tag{7}$$

where η is the viscosity coefficient of the medium and V the molar volume of the rotating sphere in milliliters. If the viscosity of the medium is changed isothermally the plot of $1/p - 1/3$ against $RT\tau/\eta$ should yield a straight line with slope proportional to $1/V$ and intercept $1/p_0 - 1/3$. For any given range of viscosity values the slope depends upon the ratio τ/V. If the aggregate and subunits have volumes V_{ag} and V_{sub}, respectively, and the corresponding polarizations are p_{ag} and p_{sub} then by Eqs. (5) and (7)

$$(1/p_{sub} - 1/p_{ag}) = (1/p_0 - 1/3)(RT\tau/\eta)(1/V_{sub} - 1/V_{ag}) \quad (8)$$

According to Eq. (8) if V_{sub} and V_{ag} are given, the magnitude of the change in fluorescence polarization following a change in aggregation is determined by the length of the fluorescence lifetime. Thus, precise measurements of the degree of dissociation of a dimer or tetramer with masses in the range of 50 to 150 KDa requires fluorophores with lifetimes of 10 to 20 nsec. If τ is shorter than about 5 nsec the changes in polarization observed will be much more influenced by the local rotations than by the overall particle rotations. It follows that the changes in the polarization of the intrinsic protein fluorescence ($\tau = 2\text{--}6$ nsec) cannot usually be relied on to give an estimate of the change in volume of the protein on dissociation into subunits. This can best be done by measurements of polarization of the protein covalently labeled with a fluorophore of appropriately long fluorescence lifetime. Experiments with small molecules [18] and with labeled macromolecules [19] yield the values of V expected from the known mass of the particles. However, in the case of most proteins the Perrin plots yield intercepts that corresponds to a value, p_{inf}, which is appreciable smaller than p_0, the one observed with the isolated fluorophore in a vitrified medium. From the fact that $p_{inf} < p_0$ we can deduce the existence of local rotations of the fluorophore, that are not appreciably influenced by the viscosities employed in the experiment, and take place independently of those of the whole particle (Fig. 3). From Eq. (6) it follows that the amplitude σ of the local rotations is determined by the relation

$$\frac{2}{3\cos^2\sigma - 1} = \frac{1/p_0 - 1/3}{1/p_{inf} - 1/3} \quad (9)$$

Equation (5) now becomes

$$(1/p - 1/3) = \left(\frac{1}{p_0} - \frac{1}{3}\right)\left(\frac{2}{3\cos^2\sigma - 1}\right)\left(1 + \frac{RT\tau}{\eta V}\right) \quad (10)$$

In conjugates of proteins with external fluorophores the observed values of

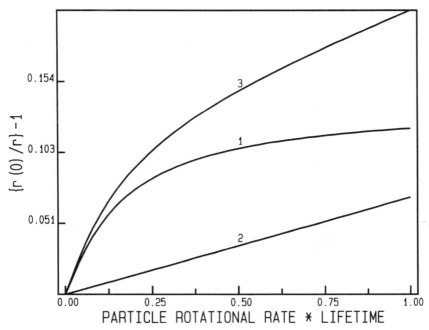

Figure 3. Separation of local and particle rotations. The stationary polarizations observed when the rotational rate of the particle [Eqs. (6) and (7)] increases progressively from negligibly small to the reciprocal of the fluorescence lifetime. 1: Local rotation. 2: Overall particle rotation. 3: Composition of both by the Soleillet rule [17] employed in Eq. (9).

p_{inf} reveal local motions of 10–30° of arc and similar values are obtained from observations of the fluorescence of tyrosine and tryptophan in globular proteins, in media of moderate viscosity in which the overall motions of the particles can be considered totally damped.

Direct real-time observations of the depolarization of the intrinsic protein fluorescence and theoretical calculations confirm the conclusions reached by the observations of polarization under stationary conditions: The local motions of the intrinsic fluorophores in proteins are performed in times that are considerably shorter than those in which the motions of the whole macromolecule become detectable.

Depolarization by the Local Fluorophore Motions

The local rotations are thermally activated motions so that their amplitude increases, and p_{inf} decreases, with temperature. In a Perrin plot of polarization data obtained by changing the ratio T/η by variation of the temperature, the slope of the straight line commonly obtained depends both on the decrease in p_{inf} with temperature as well as the whole-particle rotations, determined by the characteristic values of V and τ for the system. As a result the extrapolated value of p_{inf} is systematically larger and V calculated from the slope is systematically smaller than the values of these quantities obtained when the viscosity is changed under isothermal conditions [21]. Therefore, isothermal Perrin plots in which the viscosity, but not the temperature, is varied, must be used to determine the size of the particles [V in Eq. (10)]. The analysis of the local rotations by observations of the polarization of the fluorescence requires elimination of the concomitant rotations of the whole particle and this can be done because of the fortunate circumstance that there is a range of viscosities in which the overall particle rotations become virtually damped but the local rotations appear minimally affected; to reach this condition the viscosity of the medium must be increased just beyond the viscosity at which overall particle rotations become measurable. If the thermal depolarization of the fluorescence is examined over this range of viscosities one can obtain important information about the amplitude and rate of the local residue motions in the protein. To characterize the effect of changes in polarization of fluorescence over this thermal range of particle immobility we write the Perrin equation (6) in the form

$$\frac{RT\tau}{(r-1)} = \eta V \tag{11}$$

We expect that the change in the viscosity of the solvent with temperature may be expressed, over a suitable range of temperature, by the expression

$$\eta(T) = \eta_0 \exp[b(T - T_0)] \tag{12}$$

where η_0 is the viscosity at an arbitrary temperature T_0, which is conveniently chosen close to the midpoint of the range of temperatures employed. From the last two expressions we obtain in logarithmic form:

$$Y \equiv \ln(r-1) - \ln(RT\tau/V) = -\ln \eta_0 + b(T - T_0) \tag{13}$$

In a plot of the quantity Y against the temperature we expect to obtain a straight line with slope b corresponding to the thermal coefficient of the viscosity of the medium surrounding the fluorophore. It has been shown [22] that for small fluorophores dissolved in pure or mixed solvents the b values calculated by use of Eq. (13) are equal to the thermal coefficient of the viscosity of the solvent measured by flow. A 75% glycerol–water mixture has a flow viscosity of 2.5 P at 0° C and a thermal coefficient of the viscosity of 0.07 per degree, and this thermal coefficient is well reproduced when Eq. (13) is employed with tyrosine or tryptophan as fluorophore. For polypeptides of molecular weight greater than about 1000–2000 in 75% glycerol the contribution of overall particle rotations to the depolarization below and up to room temperature can be altogether neglected. The transparency of glycerol–water mixtures in the ultraviolet and the fact that glycerol does not seem to affect the conformation and the physiological properties of proteins, and even more complex biological systems [23], makes them ideal solvents to follow the local rotations of tryptophan and tyrosine in proteins by the use of polarization of fluorescence. Scarlata *et al.* [24] made observations on the polarization of the fluorescence from a series of natural peptides containing a single tyrosine or tryptophan. Figure 4 reproduces one of these observations: At the high viscosities attained in 75% glycerol–water at $-40°$ C, the polarizations tend to the limiting p_0 for the fluorophore, indicating the absence of appreciable rotations between excitation and emission. Between this temperature and a characteristic critical temperature T_c, the plot of Y against $T - T_0$ [Eq. (13)] has a uniform slope of 0.07 per degree, as if the frictional coefficient of rotation were determined in this range of temperatures by the solvent alone. At $T > T_c$ the slope changes abruptly over a temperature range of 3 to 6°C and becomes uniformly smaller (-0.015 to -0.055 in the different peptides) indicating that the effective frictional coefficient is no longer determined by the viscosity of the solvent. The critical temperature T_c may be interpreted as the one at which the bonds of the aromatic fluorophore with the solvent become too weak to restrict the extent of the fluorophore motions and are replaced by the constraints from nearby peptide residues. A series of observations on proteins with a single tryptophan residue (Rholam *et al.* [24]) yielded results quite comparable with those obtained with peptides, showing two, and occasionally three, abrupt changes in slope. The conclusion from the similarities of results in small peptides and in proteins is clearly that the largest influence in determining the values of b and T_c, and therefore the character of the local rotations, is that of the very few residues that surround the amino acid fluorophore.

Thermodynamics of the Local Motions

The theory of depolarization that applies successfully to the free rotation of particles is notoriously unsuited to the description of the local, restricted rotations that are observed in fluorophores belonging to a polymer or placed in a membrane. Perrin's theory of depolarization treats the problem as one of dynamics in which a gradient of the orientations of

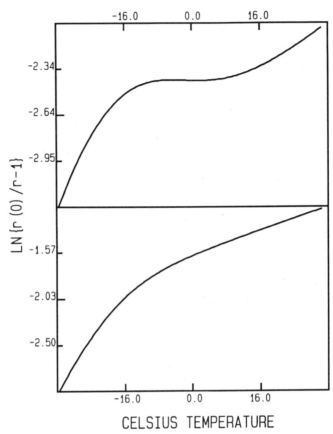

CELSIUS TEMPERATURE

Figure 4. Y-plots of local rotations [Eq. **(20)**] between the temperatures of $-32°$ and $32°C$. Parameters used in lower panel: $a_{12} = 0.30$, $a_{23} = 0.05$. At $0°$ C $f_1/f_2 = 0.3$, $f_2/f_3 = 7$. $\Delta H_{12} = -20RT$, $\Delta H_{23} = -8RT$. Upper panel: Same except $f_2/f_3 = 28$, $\Delta H_{23} = -12$. Lower panel corresponds to cases observed by Scarlata *et al.* [24]. Upper panel imitates the porphyrin–myoglobin observations [26].

the molecular oscillators, created by photoselection at the instant of light absorption, decays in time through the rotational motions of the molecules. There is, however, the possibility of considering the depolarization as resulting from a virtual exchange of orientations under conditions of thermodynamic equilibrium [26]. Before excitation, the possible molecular orientations constitute a population in equilibrium and the existence of thermally activated local rotations shows that the motions responsible are jumps between a small number of allowed fluorophore positions. For a single pair of positions between which exchanges are possible the equilibrium condition is

$$f_1 w_{12} = f_2 w_{21} \tag{14}$$

where f_1 and f_2 are the fractions of the total fluorophores in the two positions and w_{12} and w_{21} the a priori probabilities of reorientation from the direction of the first subscript to that of the second. The emission from the position of excitation generates an anisotropy of emission a_0 while emission from the alternative orientation corresponds to a smaller value $a_{12} = a_{21}$. Since there is no preferential excitation of one orientation over the other the anisotropy observed is

$$\langle a \rangle = a_0(f_1 W_{11} + s f_2 W_{22}) + a_{12}(f_1 W_{12} + f_2 W_{21}) \tag{15}$$

$W_{11} \cdots W_{21}$ of Eq. (15) are *visible probabilities*, that is probabilities that the emission will take place from the orientation of the second subscript when excitation was that of the first one. The visible probability W_{12} is dependent both on w_{12} and w_{21} as emission can take place after a number of opposing reorientations. If emission takes place after j reorientations of sense $1 \rightarrow 2$ and $j - 1$ reversals (sense $2 \rightarrow 1$), W_{12} is a mean value resulting from the sum of cases in which there is no jump ($j = 0$) between excitation and emission and an increasing number of them. Then,

$$W_{12} = \sum_{j=0}^{j=\infty} w_{12}^j w_{21}^{j-1}(1 - w_{21})\} = \frac{w_{12}(1 - w_{21})}{1 - w_{12}w_{21}} \tag{16}$$

The a priori probabilities of a reorientation during the fluorescence lifetime are $w_{12} = k_{12}\tau/(1 + k_{12}\tau)$ and $w_{21} = k_{21}\tau/(1 + k_{21}\tau)$ where k_{12} and k_{21} are the transition probabilities of reorientation per molecule per second. Introducing these relations in Eq. (16) we derive

$$W_{12} = \frac{k_{12}\tau}{1 + k_{12}\tau + k_{21}\tau}, W_{21} = \cdots \tag{17}$$

and since $f_1 k_{12} = f_2 k_{21}$,

$$f_1 W_{12} = f_2 W_{21} \tag{18}$$

The visible reorientations are therefore subjected to the same thermodynamic constraints as the a priori reorientations and this circumstance makes it possible to obtain energetic information from a study of the depolarization by local rotations. Furthermore this information relates to the structural restraints placed by the protein on the motion of the fluorophore. Making use of the conservation conditions $W_{i1} = 1 - W_{i2}$, $f_1 + f_2 = 1$ and Eqs. (15) and (18),

$$a_0 - \langle a \rangle = (a_0 - a_{12}) 2 f_1 W_{12} \tag{19}$$

Equation (15) may be generalized to include successive or simultaneous jumps between more than two orientations. If the fluorophore can adopt orientations 1, 2, and 3, with 2 as an obligatory intermediate position, Eq.

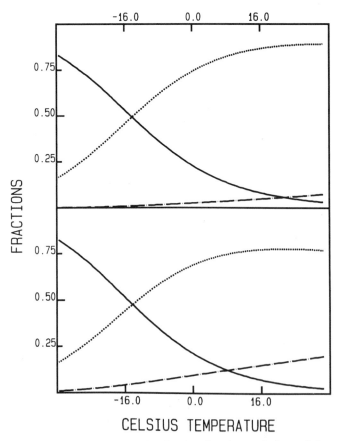

Figure 5. The variations in the fractional population of the orientations 1, 2, and 3 for the plots of Figure 4. Note that the seemingly qualitative differences between the two cases of that figure depend on very small differences in the distribution of orientations.

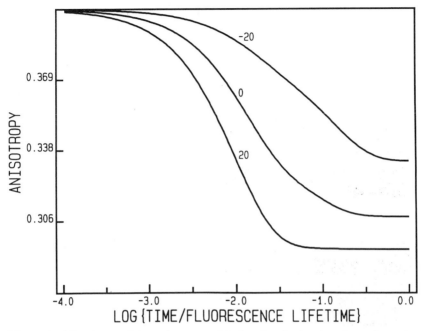

Figure 6. The decay of the anisotropy with time for the local motions shown in the lower panel of Figure 4 at 20, 0, and −20° C. Notice the differences in the final equilibrium anisotropy as well as the equilibration in a small fraction of the fluorescence lifetime.

(19) takes the form

$$a_0 - \langle a \rangle = 2(a_0 - a_{12})f_1 W_{12} + 2(a_0 - a_{23})f_2 W_{23} \qquad \textbf{(20)}$$

a_{23} is the polarization corresponding to excitation in direction 2 and emission in direction 3, W_{23} and W_{32} are given by relations like **(17)** and $f_1 + f_2 + f_3 = 1$. Equation **(20)** appears at present sufficient to interpret —qualitatively at least—the reduced number of observations on the effect of temperature on the local rotations of tryptophan residues in proteins (Figs. 5–6) as well as the observations of the limited motions in real time observed in fluorophores incorporated in membranes [25, 26]. We note that to predict the depolarization resulting from **(20)** we need to specify the thermodynamic equilibrium constants of the exchanges 1 ↔ 2 and 2 ↔ 3 at a given temperature and pressure. Further specification of the enthalpy and volume changes in the equilibria permit us to describe the expected changes in anisotropy as a function of temperature and pressure. It appears that the interpretation of the depolarization by local rotations through these ideas may permit us in the future to obtain important information of the energetics of residue interaction in proteins and therefore on the elementary protein dynamics.

Demonstration of Changes in Protein Aggregation by the Polarization of the Intrinsic Fluorescence

Local motions of the fluorophore may be expected to change with the state of aggregation when the bonds linking it to the nearby residues in the particles are weakened and this may be the case when the fluorophore is placed at the subunit boundary, or when the released monomer undergoes some structural disorganization. Because the intrinsic protein fluorophores, tyrosine or tryptophan, have too short a lifetime to register significant changes in polarization with overall protein volumes, they are well suited to reveal changes in the character of the local rotations, and thus in the strength of the bonds that link the fluorophore to the rest of the protein.

Changes in polarization observed at constant excitation wavelength can be due to changes in p_0 and in the amplitude or rate of the allowed tryptophan rotations. In many cases separation of these influences can be very difficult. The value of the limiting polarization p_0 is determined by the angle between the transition moments in absorption and emission. Ordinarily there is a unique transition direction for the fluorescence and separate directions for the transition moment directions of the different absorption bands. Excitation within each of them results in fluorescence with a different fundamental polarization p_0. The plot of p_0 against the exciting wavelength constitutes the excitation–polarization spectrum, a characteristic of each particular fluorophore and solvent [18, 27]. The lowest singlet state of tryptophan involves two distinct overlapping electronic transitions, designated L_a and L_b in Platt's [28] nomenclature. Because L_a and L_b have transition moments in absorption at a large angle to each other the value of p_0 depends on the fractional excitation of each transition. This varies rapidly with the wavelength of excitation and as a consequence the excitation–polarization spectra of indole and tryptophan shows a fine structure that is not found in most other fluorophores [29, 30]. Variations in this fine structure permit the detection of changes in the environment of the tryptophan, and therefore in protein conformation, with high sensitivity. Unfortunately, present knowledge of the exact correspondence between environmental effects and changes in the overlap of the L_a and L_b transitions, and possible changes in direction of their transition moments, does not allow a ready physical interpretation of the observed differences. These have to be evaluated in each particular case.

Although the changes in the excitation–polarization spectrum that follow changes of the state of aggregation of proteins are among the most conspicuous and regular of those observed, demonstration of their origin can be done only through their dependence on the protein concentration.

References

1. Klotz, I.M. (1989) In *Protein Function: A Practical Approach*, T.E. Creighton, ed. IRL Press, Oxford, pp. 25–54.

2. Green, N.M. (1963) *Biochem. J.* **89**, 585–591 is the classical paper on evaluation of very small dissociation constants by separate determination of the opposing rates (avidin–biotin equilibrium). DNA-repressor equilibria: Bourgeois, S. (1971) *Methods Enzymol.* **21D**, 491–500.

3. Ip, S.H.C., and Ackers, G.K. (1976) *J. Biol. Chem.* **242**, 3428–3434.

4. Behling, R.W., Yamane, T., Navon, G., and Jelinski, L.W. (1988) *Proc. Natl. Acad. Sci. U.S.A.* **85**, 6721–6725.

5. La Mar, G.M., Toi, H., Krishnamoorthi R. (1984) *J. Am. Chem. Soc.* **106**, 6397–6401.

6. Tombs, M.P., and Peacocke, A.R. (1974) *The Osmotic Pressure of Biological Macromolecules*, Part 4. Clarendon, Oxford.

7. Adair, G.S. (1928) *Proc. R. Soc. London Ser. A* **120**, 573–603.

8. Schachman, H.K. (1959) *Ultracentrifugation in Biochemistry*. Academic Press, New York.

9. Deal, W.C., Rutter, W.M., Massey, V. and van Holde, K.E. (1963) *Biochem. Biophys. Res. Commun.*, **10**, 49–54. Penniston, J.T. (1971) *Arch. Biochem. Biophys.* **142**, 322–330.

10. Leipert, T.K., Baldeschwieler, J.D., and Shirley, D.A. (1968) *Nature (London)* **220**, 907–909. For an application to an antibody–hapten binding equilibrium see Meares, C.F., Sundberg, M.W., and Baldeschwieler, J.D. (1972) *Proc. Natl. Acad. Sci. U.S.A.* **69**, 3718–3722.

11. Chandrasekhar, S. (1941) in *Stochastic Problems in Physics and Astronomy*. *Rev. Mod. Phys.* **15**, 86.

12. Bayliss, N.S., and McRae, E.G. (1954) *J. Phys. Chem.* **58**, 1002–1006. Ooshika, Y. (1954) *J. Phys. Soc. Jpn.* **9**, 594–602. Bakhshiev, G.N. (1964) *Opt. Spectrosc.* (Eng. Trans) **16**, 446–451. Jaffe, H.H. and Orchin, M. (1962) *Theory and Application of Ultraviolet Spectroscopy*. Wiley, New York, pp. 186–195. Foerster, Th. (1951) *Fluorescenz Organischer Verbindungen*, Vandenhoeck & Ruprecht, Goettingen pp. 133–140.

13. Longuet-Higgins, H.C., and Pople, J.A. (1957) *J. Chem. Phys.* **27**, 192–194.

14. Lippert, E. (1957) *Z. Elektrochem.* **61**, 962–975.

15. MacGregor, R.B., and Weber, G. (1981) *Ann. N.Y. Acad. Sci.* **366**, 140–154.

16. MacGregor, R.B., and Weber, G. (1986) *Nature (London)* **319**, 70–73.

17. Soleillet, P. (1929) *Ann. Phys. Ser X* **12**, 23–97.

18. Perrin, F. (1926) *J. Phys.* **7** 390–401. Perrin, F. (1932) *Ann. Phys. Ser. X* **12**, 169–275. This long article contains virtually all that the nonspecialist needs to know about the fluorescence of solutions.

19. Weber, G. (1953) *Adv. Protein. Chem.* **9**, 414–459.

20. Kasprszak, A., and Weber, G. (1982) *Biochemistry* **21**, 5924–5927. Gratton, E., Alcala, R.J., and Marriott, G. (1986) *Biochem. Soc. Transact.* **14**, 835–838. Ichiye, T., and Karplus, M. (1983) *Biochemistry* **22**, 2884–2893.

21. Wahl, P., and Weber, G. (1967) *J. Mol. Biol.* **30**, 371–382. Rawitch, A., and Weber, G. (1972) *J. Biol. Chem.* **245**, 680–685.

22. Weber, G., Scarlata, S., and Rholam, M. (1984) *Biochemistry* **23**, 6785–6788.

23. Gekko, K., and Timasheff, S.N. (1981) *Biochemistry* **20**, 4667–4676; **20**, 4677–4686. Na, G.C., and Timasheff, S.N. (1981) *J. Mol. Biol.* **151**, 165–178.

24. Scarlata, S., Rholam, M., and Weber, G. (1984) *Biochemistry* **23**, 6789–6792. Rholam, M., Scarlata, S., and Weber, G. (1984) *Biochemistry* **23**, 6793–6796.

25. Weber, G. (1989) *J. Phys. Chem.* **93**, 6069–6073.

26. Royer, C.A., and Alpert, B. (1984) *Chem. Phys. Lett.* **81**, 150–167.

27. Zimmerman, H., and Joop, N. (1961) *Z. Elektrochem.* **65**, 61–65, 138–142. Doerr, F., and Held, M. (1960) *Angew. Chem.* **72**, 287–294.

28. Platt, J.R. (1949) *J. Chem. Phys.* **17**, 484–495.

29. Weber, G. (1960) *Biochem. J.* **75**, 345–352.

30. Valeur, B., and Weber, G. (1977) *Photochem. Photobiol.* **25**, 441–444.

XIII

Effects of Temperature and Pressure on Molecular Associations and on Single Peptide Chain Proteins

The Role of Solvent in Molecular Equilibria

The thermodynamics and kinetics of association equilibria between two molecules—small or large—are formulated without regard to the changes in the solvent that accompany the formation of the molecular complex. The thermodynamic and kinetic parameters thus determined include the solvent contribution and this must be separately estimated when one tries to interpret the physical processes underlying the molecular association [1]. To render clear the importance of solvent participation consider the simple equilibrium $A + B \leftrightarrow AB$ where A and B may be small molecules or macromolecules. The intervention of the solvent S may be represented by means of the free energy scheme shown in Figure 1. We consider as a starting point the chemical potentials of the separate solutions of the partners, which correspond to their completely solvated states, SAS and SBS. Formation of the complex may then be imagined to take place in two successive separate steps. The first is the release of solvent from those parts of the surface of the partners that are in contact when the complex is formed. This partial desolvation corresponds to the reaction

$$SAS + SBS \rightarrow SA + BS + SS \qquad (I)$$

with a standard free energy change ΔG_d. It involves both breaking of bonds between solvent and solute and formation of new solvent–solvent (SS) bonds. The second step in the formation of the complex is the association of the partially solvated partners:

$$SA + BS \rightarrow SABS \qquad (II)$$

Its standard free energy ΔG_a arises from enthalpic and entropic contributions in the interactions of SA and BS that do not involve changes in

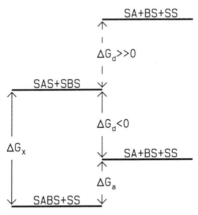

Figure 1. Diagram of the Gibbs free energy levels of a molecular complex including solvation.

solvation. The experimental free energy of association ΔG_X, derived from the dissociation constant of the complex in the usual fashion, is evidently

$$\Delta G_X = \Delta G_d + \Delta G_a \qquad (1)$$

We can think of two opposite cases: If the solvation energy of the partners is greater than that of the solvent–solvent interactions, then ΔG_d is large and positive. Formation of a complex will not take place because the partners remain in the fully solvated condition even if there could be a strong physical bond between the unsolvated partners ($\Delta G_a < 0$). Such is the case of small ions of opposite sign in water solution. At the other extreme stand the cases of weak initial solvation. The partial desolvation of the partners and their subsequent association resemble in some respects the separation of immiscible liquids in that it requires solute–solute and solvent–solvent interactions of quite different strength and character. Aromatics, as well as highly polarizable aliphatic molecules, form complexes in apolar solvents, while weakly polar aromatics do so in water and other strongly polar solvents. To the first category belong the so-called charge-transfer complexes [2] and to the second the many examples of aromatic stacking interactions encountered in biochemistry. In the charge-transfer complexes the most important contribution to the association is from solute–solute interactions, and in the stacking aromatic complexes in water this contribution derives from solvent–solvent interactions. In this latter case, in which both the solvation of the partners and the solute–solute interactions (ΔG_a) are weak, everything happens as if the solvent attempted to reduce to a minimum its area of interaction with the solute. Hence complexation may be said to result from a "solvophobic" effect. We expect that these two types of complexes that result from one dominant type of interaction, either solvent–solvent or solute–solute, will be enthalpy driven, and opposed by entropy changes in the three processes of desolvation, solvent–solvent association and desolvated partners association, that result from the inevitable enthalpy-entropy compensation. As a result the standard free energy of formation will be considerably smaller, in absolute value, than the separate entropy and enthalpy changes on complex formation.

An understanding of the thermodynamics of the complexes is greatly facilitated by an examination of the microscopic sources of the interactions. We are here primarily interested in the complexes formed by apolar molecules in polar solvents. Ideal apolar solvents are bound by dispersion forces alone while strongly polar solvents such as water, dimethyl sulfoxide, or dimethylformamide are bound by interactions among permanent dipoles, which are several times stronger than the van der Waals forces between the molecules of the apolar liquids. Association of aliphatic and aromatic molecules differs in both ΔG_d and ΔG_a because of the much larger polarizabilities of aromatics, often one order of magnitude greater than those of typical apolar aliphatic molecules with polarizabilities of only a few milliliters per mole. The importance of the mutual polarizability of aromatics in stabilizing their molecular complexes is demonstrated by the appearance, even in apolar solvents, of transient associations of unsubstituted aromatics when one of the partners is in an electronically excited state. Such complexes, excimers and exciplexes [3], owe their existence to the increased polarizability that is a characteristic of electronic excited states.

In both types of solutes, aliphatic and aromatic, the interactions with water are of the permanent dipole-induced dipole type, but the energies of water–aromatic interaction are much larger than those of water–aliphatic interactions because of the much larger polarizabilities of the former. This effect is responsible for both the larger solubility of aromatics in water and for the considerably greater energy of the van der Waals forces giving rise to ΔG_a. However, an increase in solute polarizability will similarly affect ΔG_a and ΔG_d so that a more difficult desolvation will be compensated by a correspondingly larger energy of solute interaction. Because of such compensation we expect that apolar aliphatic complexes, when suitable simple models of them become available, will prove to be similar to the aromatic complexes in most respects, including the very nearly complete enthalpy–entropy compensation that characterizes aromatic complexes in water (see below). At present the thermodynamic parameters that apply to complexes of aliphatic apolar molecules in water are unknown. The indirect calculation of these quantities starting from the free energies of transfer between polar and apolar solvents of aliphatic compounds of mixed polar and apolar character has been frequently attempted, but the results are uncertain because such calculations involve suppositions and hypothesis, the reliability of which are in doubt.

The importance of the partial desolvation in the formation of the complex is a major obstacle in any attempt to calculate the strength and character of the molecular complex SABS from structural data, and the various interatomic potentials that we described in Chapter X. A reasonable estimate can in principle be made for the unsolvated complex AB, as would actually occur in a molecular crystal in which there are only weak interactions between the near neighbors, but this would be of no great practical value unless the solvation interactions of the partners and their partial reversal on the formation of the complex in solution could be computed. Aromatic molecular complexes have experimental free energies of formation on the order of -1 to -4 kcal/mol. Formation of a stacking complex will abolish the largest part of three translational and three rotational degrees of freedom and any additional contribution to the entropy change must be attributed to other much less evident sources. In complex formation between aliphatic parts the changes in entropy due to unsolvated association are more difficult to estimate, and this difficulty becomes even greater in the case of flexible macromolecules in which many small internal adjustments among the macromolecular parts are virtually certain to occur following association.

Temperature Effects on the Association of Small Molecules

The complexes of low molecular weight formed by apolar molecules in water solution that have been analyzed in sufficient detail to separate the terms in Eq. (2) are all cases in which one or both of the partners are aromatic molecules.

$$\Delta G = \Delta H - T\Delta S \tag{2}$$

The relatively weak forces responsible for the molecular complexes predicate a small free energy of formation; consequently a rather high concentration of solutes is required to obtain appreciable association. It may not be idle to recall once more that the limited solubility of strictly apolar aliphatic compounds in water and other polar solvents, as well as the absence of sensitive methods to detect their changes in aggregation, puts a limit to the observation and measurement of the weaker complexes in which they may participate. These disadvantages deprive us of the most important experimental data in the understanding of the stability of the

Table 1. Thermodynamic Parameters of Aromatic Complexes[a]

	Solvent	ΔH	$T\Delta S$	ΔG	ΔV
1. FMN + l-tryptophan	H_2O	-7.4	-4.7	-2.7	-4
2. HMB + CQ	MCH	-8.2	-5.3	-2.9	-8

[a]ΔH, $T\Delta S$, and ΔG in kcal/mol. ΔV (change in volume on association) in ml mol^{-1}. MCH, methyl cyclohexane; FMN, flavin monomucleotide; HMB, hexamethyl benzene; CQ, tetrachloroquinone. 1: From Wilson [4]; 2: From Ewald [5].

conformations of peptide chains in solution. In aromatic compounds the disadvantages that thwart the formation and examination of aliphatic complexes are minimized: The solubility is often sufficient for the establishment of association equilibria with other aromatics and with aliphatic compounds, and the perturbation of their spectral properties on association, particularly light absorption and emission, permits the determination of the degree of dissociation of the complex with relative ease. From it a thermodynamic analysis of the equilibrium according to Eq. (2) can be readily done. Table 1 gathers two very significant examples.

In spite of the different character of the solvent the similarity of the thermodynamic parameters in the two cases shown in Table 1 is remarkable. The two cases exemplify the distinction between those complexes that are driven by solute–solute interactions (HMB + CQ) and by solvophobic effects (tryptophan and flavin). In the tetracloroquinone complexes the interaction between the solutes is strengthened by the polar character of the C—Cl bonds whereas the solvophobic effects are greatly diminished in methyl cyclohexane. There is in fact a certain symmetry in the influences that give rise to complexation in both cases: Let D and S stand for the dissolved substances that associate and solvent, respectively, and let the subscripts d and p describe the character, dispersive or polar, of the interactions in which they take part. Then the equilibria of the complexes of the two types exemplified in Table 1 may be generalized as

$$S_pS_p + D_dD_d \leftrightarrow S_pD_d + S_pD_d$$

$$S_dS_d + D_pD_p \leftrightarrow S_dD_p + S_dD_p$$

Desolvation energies, being in either case the result of dipole-induced dipole forces, need not be appreciably different. As a result of the symmetry in the energies involved we expect no great differences in the enthalpies of association of the two cases. Our almost complete ignorance of the sources of entropy decrease on association precludes a valid explanation of the similarity of the entropy–enthalpy compensation in the two cases. From these and other similar observations of the interactions of aromatic molecules in water and other solvents one can draw some simple, and apparently general, conclusions. Both enthalpy and entropy changes are negative and ΔG is in absolute value much smaller than the terms ΔH and $T\Delta S$. The major difference between the two cases shown in Table 1 is in ΔV, which increases by a factor of two in the organic solvent. A likely explanation of this difference is that the internal pressure of water (Chapter XII) is five times larger than the internal pressure of cyclohexane. As a consequence the separate aromatic partners occupy a significantly smaller volume in water [6] than in cyclohexane and the further reduction in volume, caused by the larger solute–solute interaction energy in the HMB–CQ complex, must be correspondingly greater than in the complex of tryptophan and isoalloxazine. The realization that complexes between aromatic molecules are most conspicuous in water solution, and often break down on addition of a somewhat less polar component, such as ethanol, has given rise to the impression that interactions in water have some specific character absent in other solvents, but weaker aromatic associations have been observed in other polar solvents such as dimethylformamide or dimethyl sulfoxide. The explanation of the unique position of water is probably very simple: Although the molecular interactions in water are of the same character as in other solvents with permanent dipoles, the dipole density of water is twice as great as those of the most comparable of the polar solvents, methanol and formamide.

The need for a correct assessment of the solvent contribution to the stability of the complex becomes particularly evident when one tries to determine the solvent contribution to the effects that follow changes in temperature or pressure in cases that involve solvophobic association. As the temperature decreases, dipole association increases and partially neutralized molecular aggregates form. As a result, at low temperature a polar solvent loses part of its polar character and can display some properties of a nonpolar solvent. It then becomes capable of separating the partners of molecular complexes that are stable at higher temperatures. The simplest phenomenological description of the effect of changes in the solvent on the complex is one in which the solvent is assumed to exist in two forms in equilibrium, S and S* that coexist in proportions given by the free energy change $\Delta G(S \rightarrow S^*)$, proportions that may be altered by changes in temperature and pressure. In attempting to evaluate the importance of changes in the solvent to the energy of desolvation a reasonable assumption would be that the change in ΔG_d is determined by the proportions of the solvent forms in equilibrium, S and S*. The free energy of desolvation is now an average given by

$$\langle \Delta G \rangle_d = f(S)\Delta G_d + f(S^*)\Delta G_d^* \qquad (3)$$

with ΔG_d, ΔG_d^* as the free energies of desolvation;

$$f(S) = \left[1 + \exp(-\Delta G(S \rightarrow S^*)/RT)\right]^{-1}$$

$$f(S^*) = 1 - f(S) \tag{4}$$

$$\langle \Delta G_X \rangle = \Delta G_a + \langle \Delta G_d \rangle \tag{5}$$

To compute the effects caused by changes in temperature we need to specify the values of $\Delta G(S \rightarrow S^*)$ and ΔG_a at a fixed temperature and the corresponding standard enthalpy changes $\Delta H(S - S^*)$ and ΔH_a. It is particularly instructive to evaluate the importance of the anhydrous association parameters, ΔG_a and ΔH_a, by varying them, while keeping all other thermodynamic parameters constant. Figure 2 shows the result of a comparison between an association without appreciable affinity of the anhydrous moieties, a case that we expect to approximate the association of aliphatic apolar molecules, and an association to which we attribute the standard free energy and enthalpy changes that have been observed in the stacking complexes of aromatic moieties. We assume that solvation by the polar solvent increases as the temperature decreases, as a result of the self-neutralization of the solvent dipoles, in exactly the same way in both cases. Figure 2 shows that differences in ΔG_a alone are sufficient to produce a qualitative difference as regards the temperature dependence off the association: In the temperature window above zero, heating results in association of the complex with the negligible enthalpy ΔH_a but produces the opposite effect in the complex with the larger enthalpy. At the lower temperatures heating results in dissociation in both cases. Thus the effects are not only quantitatively different, but, depending on the temperature window accessible to observation, they may appear to be directly opposite as regards the sign of the temperature coefficient. Figure 2 demonstrate the misleading simplification that is involved in the statement that "a decrease in temperature uniformly promotes dissociation of apolar complexes." It seems important to add that when one or both of the associating dissolved molecules is a flexible polymer, changes in the number of degrees of freedom (heat content) with temperature and pressure are to be expected, and their separation from those changes owing to solvent alone is made that much more difficult.

Pressure Effects on the Association of Small Molecules

The effects of pressure on the distance between interacting atoms or molecules depend on the character of the bonds that link them. By an examination of the compressibility of pure liquids it is possible to crudely

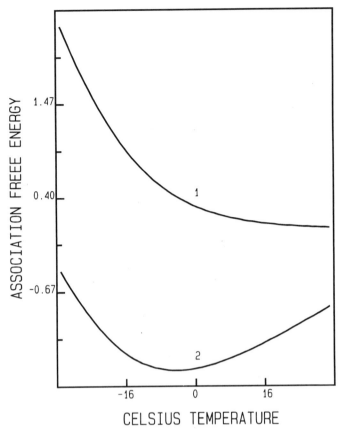

Figure 2. Effect of temperature on the association equilibria of complexes 1 and 2 that differ in the enthalpy of association, in a solvent that is present in forms S and S*. The free energies quoted are those valid at 0°C. The energy parameters, all in RT units, are $\Delta G(S \rightarrow S^*) = 3$, $\Delta H(S \rightarrow S^*) = -20$; $\Delta G_d(S) = 0$, $\Delta G_d(S^*) = 6$; $\Delta G_a(1) = 0$, $\Delta G_a(2) = -2$; $\Delta H(1) = 0$, $\Delta H(2) = -11$.

relate the resistance to compression of the various types of intermolecular bonds. As expected the weaker the interactions the more compressible is the liquid, the gradation extending from the more compressible (dispersion forces) to the least compressible (interactions between permanent charges). Table 2 gives some relevant liquid compressibilities.

Unfortunately it is not possible at this time to fruitfully relate the compressibilities of bonds to the expected volume changes on reaction,

Table 2. Compressibilities of Some Pure Liquids, Expressed as the Volume of 1 ml (at 1 bar) at 1 kbar (V_1), and the Corresponding Linear Contraction (l_1)[a]

	V_1	$l_1 = 1 - V_1^{1/3}$
Hexane	0.906	0.0341
Benzene	0.924	0.0260
Ethanol	0.931	0.0235
Water	0.957	0.0145

[a]From ref. [7].

beyond the statement that equilibrium favors the formation of the most stable and shorter of them. We then expect that *if there are no restrictions to the approach of the partners to reach the most favorable interaction*, decrease in volume with complexation will be the general rule. Exceptions to this rule may be expected when the approach of the partners is opposed by resistance to deformation from a covalent bond skeleton. The expectation just formulated appears confirmed by the regular decreases in volume observed on association, not only in complexes between aromatic molecules, but also in those formed by I_2 with diethyl ether and diethyl sulfide [8].

The restraining influence of covalent bonds appears clearly in a case studied by Visser *et al.* [9]. They examined the effects of pressure on the intramolecular complexes of flavinyl tryptophan methyl esters in which the interacting tryptophan and isoalloxazine moieties were separated by either three or five methylene links. With five links, rotation about the intervening C—C bonds permits virtually unrestricted approach of the flavin and indole rings, but with only three links this approach is severely limited. With the five-methylene separation (free approach) the standard volume change on formation of the complex was -4.7 ml mol^{-1} and with the three-methylene separation (restricted approach) it was only -1.8 ml mol^{-1}. Clearly, cases must exist in which the restriction to the approach of the partners is sufficiently strong to result in an increase in volume on formation of the complex. Torgerson *et al.* [10] found that such was the case with the inclusion complexes of poly-β-cyclodextrin [11] with two naphthalenic compounds: 1-anilino-8-naphthalene sulfonate and 2-dimetheylamino-6-propionyl naphthalene, where the standard change in volume on complex formation at atmospheric pressure was in both cases $\simeq 9$ ml mol^{-1}. In the inclusion complexes of the cyclodextrins the intermolecular distances in the dextrin–water or dextrin–aromatic associations are unchanged, as a result of the virtual incompressibility of the covalent bonds linking the atoms that form the cyclodextrin cup within the range of pressures used (1–10 kbar). Following a rise in pressure, minimum volume of the system is realized if the least compressible partner (water) occupies the cyclodextrin cup replacing the more compressible partner (aromatic molecule). By a simple application of Le Chatelier's principle we then expect dissociation of the inclusion complex on raising the pressure. The compressibility of the included aromatic molecule was derived from the equilibrium data and shown to approach that of a typical aromatic fluid such as toluene. Apart from the inclusion complexes of the aromatics no case has been described of a well-defined complex destabilized by increase in pressure.

In comparing the effects of temperature and of pressure on molecular associations we can readily see that the interpretation of the former in terms of molecular events is a more complex task than the latter: An increase in temperature produces both an increase in the internal energy of the system and of its volume. Increase in pressure produces only a decrease in volume, so that in principle, and in practice in homogeneous systems, one could isolate the effects of change in energy in the system from those caused by change in volume by means of observations at various temperatures but at constant volume [12]. However, in the case of molecules in solution the isolation of the effects caused by the increase in thermal energy from those that are dependent only on the concomitant increase in volume cannot be simply done by keeping the same overall volume of the system. In dilute solutions maintenance of the same volume as the temperature is increased practically corresponds to maintenance of the solvent volume. Accordingly, it does not entail the neutralization of the volume change with temperature for each equilibrium of the components in solution, but results in additional equilibrium perturbations that may be just as difficult to analyze as those linked with an unrestricted change in volume.

Association of Peptide Chains

Our limited knowledge of the properties of molecular complexes and our inability to analyze quantitatively the simpler molecular complexes of aliphatic molecules imply an even more primitive state of affairs as regards protein associations. In proteins the association of large parts of the covalently linked chain among themselves, or the association of separated peptide chains into specific aggregates held by noncovalent forces, results from multiple interlocked equilibria. The effects of temperature and pressure on the various equilibria involved in the overall effect of aggregation may be expected to be far from simple and application of Eq. (2) to these cases can hardly be expected to be very illuminating. Consider in this respect the equilibrium between the partially extended and compact forms of a protein made up of two portions of the peptide chain P and P' symbolically depicted as

$$PP' \leftrightarrow P - P'$$

Here PP' stands for the compact form and P − P' for the form in which two seemingly compact globules are separated by a stretch of extended peptide chain. Formulated as a simple equilibrium between two forms of a polymer we will expect it to obey the general expression

$$\frac{[PP']}{[P - P']} = \exp\left(\frac{\Delta G}{RT}\right) = \exp\left(\frac{\Delta H}{RT} - \frac{\Delta S}{R}\right) \qquad (6)$$

At the characteristic temperature T_c, $[PP']/[P - P'] = 1$ and

$$\Delta H = T_c \Delta S \qquad (7)$$

so that Eq. (6) may be written

$$\ln\left(\frac{[PP']}{[P - P]}\right) = \left(\frac{\Delta H}{RT}\right)\left\{1 - \left(\frac{T}{T_c}\right)\right\} \qquad (8)$$

Equation (8) cannot be expected to be valid except in a restricted temperature interval about T_c. Over a larger temperature interval the forms of the chain present will conceivably differ from those that are typically present at T_c: As the temperature rises above T_c, bonds resulting from internal contacts of each chain are broken and both the standard enthalpy ΔH, and the entropy of the chain increase. The effects are opposite if the temperature is decreased below T_c. In either case, structurally and functionally P and P' are no longer the same molecules that were present at T_c. We have here another serious shortcoming of the concept of chemical potential: Its rapid variation with temperature in the more flexible molecules renders virtually useless any attempt to extract exact information from equations such as (8). Their interpretation can provide only a semiquantitative picture of the dependence of the equilibrium with temperature. Additionally, the simple monomolecular equilibrium between compact and extended forms of a chain just discussed may be anything but simple: In agreement with actual experimental observation, e.g., in serum albumin [13], we know that the extended P − P' form can be brought about by protonation of the charged carboxyl groups, thus creating an electrostatic repulsion between the positively charged lysines and arginines in the P and P' moieties. At pH values at which the two forms, compact and extended, are in equilibrium this will be more properly described by the following scheme:

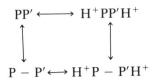

We expect that the forms present at equilibrium in detectable amounts will be PP′ (compact) and H^+P—$P′H^+$ (extended) while the other two forms are transient fluctuations that permit the establishment of the equilibrium. What was expected to be a simple first-order equilibrium is in reality a more complex one involving both first- and second-order reactions. The effect of the temperature on the equilibrium will result from a balance not from one process and its exact reversal but between two independent processes: The temperature dependence of the protonation of the groups responsible for the electrostatic repulsion of the P and P′ parts will follow the enthalpy of ionization and therefore the thermal coefficient of the pK of the protonated groups, dpK/dT, while the return of the extended, unprotonated form $P - P′$ to the compact form PP′ will be determined mainly by the entropy of the association process.

The enthalpy of ionization of the protonated groups being positive we expect that a rise in temperature will favor the deprotonated, compact form PP′ but at the same time the increase in entropy of the flexible connecting link will favor the persistence of the extended form $P - P′$. The overall thermal coefficient will depend therefore on the relative effects of temperature on these two opposing, and very different, processes. In serum albumin, extension of the flexible links between compact portions of the molecule takes place in the pH interval 4–2.5 and because of the very small temperature dependence of groups of this low pK value we expect that raising the temperature will favor the extended form. This expectation is confirmed by experiment.

An even more complex temperature behavior may be expected in the case of oligomeric proteins in which salt linkages make appreciable contributions to the stability of the aggregate. Intersubunit salt linkages have been described in hemoglobin and lactate dehydrogenase among other oligomeric proteins. If AA stands for a dimer in which the subunits are linked by two or more electrostatic linkages the simple dissociation equilibrium

$$AA \longleftrightarrow A + A$$

can be more properly formulated as

$$A\begin{vmatrix} COO^- - H_3^+N \\ NH_3^+ - {}^-OOC \end{vmatrix} A \longleftrightarrow 2A \begin{vmatrix} COO^-(H_2O) \\ NH_3^+(H_2O) \end{vmatrix}$$

The unhydrated salt linkages in the dimer will become fully hydrated, ionized groups on dissociation. The contact of the charged groups with the water dipoles gives rise to a decrease in volume by electrostriction. At neutral pH we expect that the disruption of the salt linkages must be caused by the transient protonation of the carboxyl groups and a rise in temperature will have only a minor effect on this cause of dissociation. On the other hand the hydration of the separated charges will be stabilized by a decrease in temperature so that dissociation will be favored at low temperature, in agreement with the observations on association equilibria of a number of protein aggregates at atmospheric pressure and of dimers and tetramer proteins under hydrostatic pressure (Chapter XV).

In general the effects of temperature on protein equilibria are difficult both to follow and to interpret. As deduced from the application of the various spectroscopic methods, small changes in temperature have important effects on the protein conformation, whereas the application of pressure does not seem to produce comparable effects. This difference arises because the rise in temperature results in changes in volume of the system as well as increases in the entropy (disorder) of the system. The latter changes may be generally disregarded for moderate increases in pressure.

The Effect of Pressure on Proteins Made up of a Single Peptide Chain

The effects of hydrostatic pressure in the range of 1 atmosphere to 10 kbar have been studied in about a dozen proteins made up of a single peptide chain since the original observations of Suzuki *et al.* [14] on the reversible aggregation of serum albumin subjected to a pressure of several kilobar. The pressure limit is dictated by the freezing of water at room temperature, which occurs at about 12 kbar. The methods of study have been virtually limited to optical spectroscopy, including Rayleigh scattering, absorption and emission of light by the aromatic amino acids [15], subsequently Raman scattering [16], and recently NMR [17]. The absorption and emission of light, particularly the latter, can be accurately measured at protein concentrations much lower than Raman scattering or NMR, but these two latter methods can yield information on features of the protein at the atomic resolution level. Use of these techniques indicates that, at neutral pH, hydrostatic pressures below some 4 kbar produce only minor spectroscopic changes. Beyond this pressure, characteristic spectral changes take place in all the proteins studied: In absorption there is an increase in absorptivity of the tryptophan at the longer wavelengths (> 290 nm), similar to that observed in difference spectral measurements after addition of denaturing agents [18].

In metmyoglobin changes in the absorption spectrum of heme, characteristic of new axial ligands, are observed [19]. In fluorescence there are equally general effects: The emission of tryptophan shifts to the red, sometimes by as much as 20 nm [20] and generally undergoes some decrease in yield. These spectral changes are those that may be expected on exposure of the tryptophan to a very polar environment, which we may surmise is provided by the water molecules. The spectral changes become significant at pressures higher than about 5 kbar and in most of the proteins they reach stable values in the range of 8–10 kbar. Li *et al.* [20] observed that chymotrypsin and lysozyme become capable of binding, respectively, one and two molecules of anilinonaphthalene sulfonate in the range of pressures in which these changes occur, indicating the abrupt appearance of a new conformation with altered binding properties. In most proteins the pressure changes are completely reversed on decompression but Visser *et al.* [21] found in the flavodoxins that reversibility, though very imperfect at pH 7, was greatly improved at pH 5. Because a plateau in the spectral effects is reached in the pressure range of 8–11 kbar it has been concluded that pressure induces a transition between two forms of the protein: "native" at atmospheric pressure and "pressure denatured" at the highest pressure values. The ratio of the concentrations of the two forms is

$$\frac{[D]}{[N]} = K_P; \qquad \frac{K_P}{K_{atm}} = \exp\left(\frac{p\Delta V}{RT}\right) \tag{9}$$

where K_P is the equilibrium constant at pressure p and ΔV is the standard volume change in the transformation of N into D. Values of ΔV of -30 to -100 ml mol^{-1} have been calculated from the spectral data [8, 15]. These changes are a small fraction, on the order of 0.25%, of the molar protein volume. From dilatometric experiments it has been concluded [22] that folding of a peptide chain into a globular protein should result in an increase in volume of about 2%, and Richards has observed that the amino acid residues in globular proteins occupy larger volumes (by about 5%) than they do in crystals of the amino acids [24]. Evidently, pressure denaturation does not lead to more than a very small change in the protein volume and one might easily assume that this is the result of

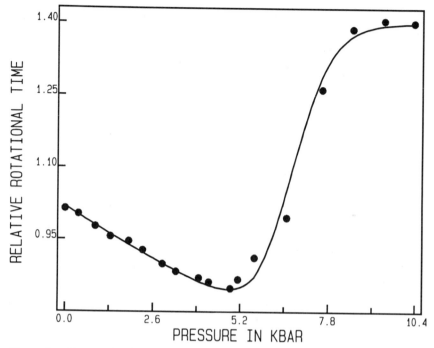

Figure 3. Changes in the relative rotational diffusion of lysozyme by compression in the range 1–10 kbar. The curve is the sum of a volume reduction showing a compressibility of 0.0402 kbar^{-1}, and pressure conversion to the "denatured" form [Eq. (9)] $[D]/[N] = \exp[(p - p_{1/2})\Delta V/RT]$ with $p_{1/2} = 6.5$ kbar, $\Delta V = 40$ ml mol^{-1}. Redrawn from ref. [23].

reduction or disappearance of small free volumes or packing defects distributed through the structure, the energetic debt for their collapse being paid by the work of compression, $p\Delta V$. In that case we would expect that the hydrodynamic volume of the protein would remain practically the same on compression, but a very different conclusion is reached by measurements of the rotational relaxation time of lysozyme under increasing high pressure [23], shown in Figure 3. In the pressure region below ~ 5 kbar the hydrodynamic volume decreases almost linearly to a minimum of about 90% and then, in the pressure region in which other spectroscopic changes take place (5.7–7 kbar), the hydrodynamic volume increases rapidly and reaches a stable value that is some 60% larger than that at atmospheric pressure. As compression results in a diminished volume of protein + solvent, but with a greatly increased hydrodynamic volume, we must explain the mechanism by which bonds in the protein interior are sufficiently weakened by application of pressure to allow invasion by the solvent with partial disappearance of the small packing voids existing in the native protein.

It seems reasonable to expect that increasing pressure at constant temperature decreases the distances between the amino acid residues with a consequent increase in the Born repulsion. Rotation about bonds that link the amino acid residues to the backbone will work to reduce the repulsion at the expense of creating larger free volumes that the solvent can now comfortably occupy. The result will be a swollen molecule in which many short-range amino acid interactions are replaced by interactions with water molecules interspersed in the structure. This model of the pressure-denatured protein resembles in most experimental respects, including the binding of anilinonaphthalene sulfonate [20, 26], the first state of protein folding reached after the denaturing agent (urea or guanidine) is removed by dilution, and which Ptytsin has named the molten globule [25]. Under high pressure the protein seems to be penetrated by water in the same way as it is penetrated by urea or guanidine, added at high concentrations, at atmospheric pressure.

For the purpose of analyzing the effects of pressure on protein associations the most important conclusion to be derived from observations of the pressure effects on single chain proteins is that conspicuous changes in the conformation of peptide chains do not seem to take place at pressures under 4 to 5 kbar.

References

1. For various views on this topic see Kauzmann, W. (1959) *Adv. Protein. Chem.* **14**, 1–63, which contains the original formulation of the importance of the solvophobic effects in the structural stability of globular proteins. Tanford, C. (1973) *The Hydrophobic Effect.* Wiley, New York. Ben-Naim, A. (1980) *Hydrophobic Interactions.* Plenum, New York. Nemethy, G., Peer, W.J., and Scheraga, H.A. (1981) *Annu. Rev. Biophys. Bioeng.* **10**, 459–497. Pratt, L.R., and Chandler, D. (1977) *J. Chem. Phys.* **67**, 3683–3704.

2. Mulliken, R.S. (1952) *J. Am. Chem. Soc.* **74**, 811–824. Briegleb, G., and Czekalla, J. (1960) *Angew. Chem.* **72**, 401–413. Deranleau, D., and Schwytzer, R. (1969) *Biochemistry* **9**, 126–134. Most of the interest in charge-transfer complexes has centered on their excited state properties and scant attention has been paid to the determination of the ground-state thermodynamic parameters.

3. Birks, J.B. (1967) *Nature (London)* **214**, 1187–1190. Birks, J.B. (1970) *Prog. React. Kinet.* **5**, 181–272.

4. Wilson, E.J. (1966) *Biochemistry* **5**, 1351–1359.

5. Ewald, E.H. (1968) *Trans. Faraday. Soc.* **64**, 733–743.

6. Masterton, W.L. (1954) *J. Chem. Phys.* **22**, 1830–1833.

7. Bridgman, P.W. (1931) *The Physics of High Pressure*, Chapter V. Dover, New York.

8. Weber, G., and Drickamer, H.G. (1983) *Q. Rev. Biophys.* **16**, 89–112. Williams, R.K. (1981) *J. Phys. Chem.* **85**, 1795–1799. Ewald [5]. I_2–ethers complexes: Sawamura, S., Taniguchi, Y., and Suzuki, K. (1979) *Bull. Chem. Soc. Jpn.* **52**, 281–283, 284–286.

9. Visser, A.J.W.G., Li, T.M., Drickamer, H.G., and Weber G. (1977) *Biochemistry* **16**, 4883–4886.

10. Torgerson, P.M., Drickamer, H.G., and Weber G. (1979) *Biochemistry* **18**, 3079–3083.

11. Harada, A., Furue, M., and Nozakura, S. (1977) *Macromolecules* **10**, 676–681. The inclusion complexes of the cyclodextrins were first described by Cramer, F., Saenger, W., and Spatz, H.Ch. (1967) *J. Am. Chem. Soc.* **89**, 14–20.

12. Jonas, J., DeFries, T., and Wilbur, D.J. (1976) *J. Chem. Phys.* **65**, 582–588.

13. Yang, J.T., and Foster, J.F. (1957) *J. Am. Chem. Soc.* **76**, 1588–1595. Weber, G., and Young, L.B. (1964) *J. Biol. Chem.* **239**, 1415–1423.

14. Suzuki, K., Miyosawa, Y., and Suzuki, C. (1963) *Arch. Biochem. Biophys.* **101**, 225–228.

15. Heremans, K. (1982) *Annu. Rev. Biophys. Bioeng.* **11**, 1–21.

16. Heremans, K., and Wong, P.T.T. (1985) *Chem. Phys. Lett.* **118**, 101–104.

17. Jonas, J. (1982) *Science* **216**, 1169–1174.

18. Brandts, J.F., Olivera, R.J., and Westort, C. (1970) *Biochemistry* **9**, 1038–1047. Hawley, S.A. (1971) *Biochemistry* **10**, 2436–2442.

19. Zipp, A., and Kauzmann, W. (1973) Biochemistry **12**, 4217–4228.

20. Li, T.M., Hook, J.W. III, Drickamer, H.G., and Weber, G. (1976) *Biochemistry* **15**, 5571–5580.

21. Visser, A.J.W.G., Li, T.M., Drickamer, H.G., and Weber, G. (1977) *Biochemistry* **16**, 4879–4881.

22. Zamyatnin, A.A. (1984) *Annu. Rev. Biophys. Bioeng.* **13**, 145–165.

23. Chrysomallis, G., Torgerson, P.M., Drickamer, H.G., and Weber, G. (1981) *Biochemistry* **20**, 3955–3959.

24. Richards, F.M. (1979) *Carlberg Res. Commun.* **44**, 47–63.

25. Ptytsin, O.P. (1987) *J. Protein. Chem.* **6**, 273–293.

26. Ptytsin, O.P., Pain, R.H., Semisotnov, G.V., Zerovnik, E., and Razgulyaev, O.I. (1990) *FEBS Lett.* **262**, 20–24.

XIV

Dissociation of Protein Dimers

The formation of complexes of specific proteins is a central event in the development of the structural and functional characteristics of living systems. However, the energetics and dynamics of such complexes are very poorly understood and only some of the structural aspects can be considered as reasonably well defined. Here, as in all other cases of molecular interactions, a detailed examination of the simpler cases may serve in deriving rules and insights that are useful in the analysis of the more complex ones. Evidently the simplest molecular aggregate is a dimer protein made up of identical subunits, and our aim should be to clearly understand the energetics, dynamics, and structural features that determine the equilibrium of the dimer and the subunits.

Classical Dimer–Monomer Equilibria

According to classical chemical thermodynamics the equilibrium of dimer and monomers admits a single dissociation constant K related to α, the degree of dissociation of the dimer by the expression

$$K = \frac{4\alpha^2 C}{1 - \alpha} \qquad (1)$$

where C is the total protein concentration as dimer. In logarithmic form:

$$\log \frac{K}{4C} = \log\left[\frac{\alpha^2}{(1 - \alpha)}\right] \qquad (2)$$

The existence of a unique dissociation constant can be readily recognized in a plot of the degree of dissociation, α against $\log C$, which I shall call a dilution curve. As indicated by Eq. (2) the characteristic span of concentrations of the dilution curve between the values of $\alpha = 0.1$ and $\alpha = 0.9$ is 2.86 decimal logarithmic units.

Dissociation of Neurophysin II

As a result of the difficulties mentioned in discussing the methods of investigation of protein dissociation we find few reasonably complete data regarding the dissociation of protein dimers. What may be the simplest example of dimer dissociation, and one of the few well-studied cases, is that of the neurohypophysial protein hormone neurophysin II [1], which acts physiologically as a carrier of the pituitary hormones vasopressin and ocytocin. Rholam and Nicolas [2] applied the classical hydrodynamic methods of sedimentation and diffusion to it and also performed experiments of fluorescence depolarization of the dansyl conjugates. They were able to experimentally check the validity of Eq. (1) and to obtain the values of K for the dimer, both free and saturated with the physiological ligand octosin. From their studies of the free energy of dissociation of the monomers in the absence and presence of ligands and from the titration curve of the equilibrium of the dimer with ocytocin it is possible to derive a complete scheme for the free energy couplings between ligand binding and monomer association. The scheme of Figure 1, taken from Scarlata and Royer [3], shows that only a small, barely significant, increase in free energy of dimer association takes place when the first molecule of ocytocin is bound, whereas a very significant increase in free energy of association takes place on binding a second molecule of ocytocin. The free energy couplings between ligand binding and subunit association are clearly second order. A purely second-order coupling between the free energies of ligand binding and dimer association implies that any significant changes in conformation of the dimer occur following the binding of the second ligand, so that a two-state theory would suit the present case far better than it does hemoglobin. However, if the small difference of 0.45 kcal between $\Delta G(0)$ and $\Delta G(1)$ were regarded as significant we would be forced to conclude that a fraction of the changes in the free energy of dimer association follow already the binding of the first ligand, that the coupling is not solely of second order but with some admixture of intermediate order, and that in describing the effects of ligand binding on monomer association it is necessary to postulate a number of states at least equal to the number of states of ligation, three in our case.

An analysis of the local motions of tyrosine in neurophysin has been carried out employing observations of the polarization of the intrinsic fluorescence as described in Chapter XII. The Y-plots for neurophysin

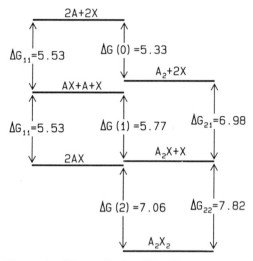

Figure 1. Neurophysin II: Free energy scheme showing the second-order coupling between the binding of ligand and subunit association. From ref. [3].

monomer and dimer consist of the typical two slopes observed in smaller peptides, with the parameters $T_c = -12°\,C$, $b(S) = 0.07$, and $b(U) = 0.035$. Thus the local motions of the unique tyrosine residue in each monomer are unaffected by association into the dimer. A different situation is obtained for the liganded dimer: Here three slopes [$b(S) = 0.07$, $b_1(U) = 0.053$, $b_2(U) = 0.034$] are found limited by two values of T_c, $-26°$ and $+5°\,C$. The intermediate slope [3] has the characteristics that may be expected from two different tyrosine populations with similar b parameter and different T_c values. From these observations it has been concluded [2, 3] that an asymmetry in the spectroscopic properties of the tyrosines as well as in their local motions is induced by the binding of the specific ligands. The asymmetry observed on ligation exemplifies the kind of microscopic information that may be derived from an analysis of local motions by the use of the polarization of the intrinsic fluorescence. The monomer–dimer equilibrium of neurophysin II is a good example of the association between two very rigid structures that show only small changes in conformation on change in the state of aggregation. This finding is not surprising in view of the very high cross-linking (seven sulfhydrils per monomer) that renders most unlikely any substantial internal motions on dissociation of the dimer. A significant difference is noticeable on comparing the dissociation equilibrium of neurophysin II with that of another dimer protein, yeast enolase.

Dissociation of Yeast Enolase: Conformational Drift of the Isolated Subunits

This dimer dissociation has also been studied by both hydrodynamic and spectroscopic methods [4]. The sedimentation observations show that dilution of the protein to concentrations of several micromolar, at low ionic strength, is not followed by conspicuous dissociation although a variation of the sedimentation constant with the rotor speed of the centrifuge was reported. As discussed in Chapter XV this phenomenon is characteristic of an incipient pressure-induced dissociation of the protein. At higher ionic strengths (1 M KCl) the dissociation of the protein into dimers can be readily detected by measurements of sedimentation and diffusion and by sedimentation equilibrium. Spectroscopic observations of the protein at the same ionic strength indicate large changes in the emission spectrum characteristic of an increase in the polarity of the tryptophan environment. Observations of tryptophan fluorescence at low ionic strength can be carried out down to 10 nM concentrations [5], and permit one to determine the complete dilution curve. From this study the conclusion is reached that at low ionic strength the dimer dissociation constant is in the vicinity of 0.1 μM and that it increases by more than two orders of magnitude at the high salt concentrations. More important, the dilution curve is evidently anomalous, with a logarithmic span of about 1.6 log units instead of the 2.86 units expected from Eq. (2). The shortened span can also be seen on recording the effect of dilution on the polarization of the fluorescence from a dansyl conjugate. Use of this conjugate permits a test of the overall reversibility of the dissociation: If the conjugate is diluted to virtually complete dissociation and unlabeled protein is added at a concentration of about 50 times the dissociation constant, the polarization of the fluorescence from the dansyl fluorophore increases slowly and finally stabilizes at a value corresponding to the undissociated enolase conjugate.

Dependence of the Chemical Potentials on Extent of Reaction

There is at least one previous report in the literature regarding the dissociation of a dimer, malate dehydrogenase, in which a reduced span of the dilution curve has been observed [6]. Recently Silva *et al.* [7] observed that the dilution curve of the ARC protein, a dimer of 14,000 molecular weight, has a similar shortened logarithmic span, 1.4 log units. In attempting to explain the reduced span of the dilution curve of enolase Xu and

Weber [5] were led to consider that the increase in dissociation constant with dilution arises in a possible change in conformation, and therefore in chemical potential of the monomers when they become separated from each other, a possibility that we have already mentioned in Chapter II. If the changes in conformation following the dissociation of the dimer into monomers are very fast in comparison with the time for an association–dissociation cycle (AD cycle), monomers and dimers will have virtually unique conformations independent of the degree of dissociation; we may then expect a single dissociation constant and a dilution curve spanning 2.86 log units between the degrees of dissociation of 10 and 90%.

The time for an AD cycle increases regularly with dilution and the increase is, in the classical picture, exclusively due to the lengthened lifetime of the monomer. At small degrees of dissociation the relatively short lifetime of the monomers permits only minimal changes in conformation following the dissociation of the dimers, and the dimers resulting from subsequent monomer association will resemble in structure and properties those that exist in the most concentrated solutions, at which there is a vanishingly small degree of dissociation. At large dissociation the greatly increased lifetime of the monomers permits them to undergo significant conformal changes and the dimers resulting from the association of such monomers will differ in structure and properties from the dimers present at vanishingly small dissociation. The lifetime of the dimer will determine whether the more conspicuous changes in conformation of the monomers are reversed during it or otherwise. The increasingly changed conformation of the monomers as the lifetime of the free monomer increases entails a partial loss of the affinity for each other as the degree of dissociation increases. If the lifetime of the dimer is insufficient for the complete reversal of the changes on which the loss of affinity depends, the dissociation constant will steadily increase with degree of dissociation and will result in a decreased logarithmic span of the dilution curve. In these cases the chemical potentials of the forms in equilibrium cannot be regarded as constant quantities but as dependent on the extent of reaction. Indeed, we may regard the classical notion of the independence of the chemical potentials on the extent of reaction as a limiting case that we often find in practice to hold by reason of the limited precision of our experimental measurements.

The observation of the dependence of the chemical potential on the extent of reaction in the more complex, flexible molecules should come to us as no surprise. In small molecules changes in conformation are notoriously fast, virtually always in the range of a few to a few hundred picoseconds, whereas the time for an AD cycle of a molecular complex with free energy of association of 2–4 kcal is in the range of nanoseconds to microseconds. In the dissociation of large flexible polymers such as the proteins the AD cycle is much longer and it is only in them that undeniable, *slow* changes in conformation have been observed. The evidence obtained from the study of the effects of pressure on the oligomeric proteins (see Chapter XV) is unequivocal in the indication that the subunits of an oligomeric protein undergo changes in the original conformation on separation and that the reversible changes that follow cannot be described as a simple isomerization but have a more complex origin. For this reason we shall refer to the reversible changes in properties of an oligomeric protein on dissociation by what we believe to be its cause: the *conformational drift* of the isolated subunits. Aggregation of imperfectly folded subunits, followed by a first-order process of reactivation, that must be attributed to the regeneration of the native structure in the aggregate have been observed in dimers [8] and in tetramers [9].

Phenomenological Description of Dimer–Monomer Equilibria Subjected to Conformational Drift

To fit the experimental facts we need to replace Eq. (1) by one that describes the dependence of K upon C in the general case. We may adopt one of two different points of view: We may consider the dimers and monomers as forming, at a given degree of dissociation, a *homogeneous* population with a standard free energy of association $\Delta G(\alpha)$ intermediate between $\Delta G(0)$, operative when $\alpha \to 0$, and $\Delta G(1)$, operative when $\alpha \to 1$. Alternatively we can envision a *heterogeneous* population that results from the mixing of monomers of two different kinds: M that exists in equilibrium with dimers at the lowest degrees of dissociation, and M*, a conformationally derived species that is predominant in the very dilute solutions in which dissociation is almost complete. The dimers resulting from association of these monomers are of the types MM, M*M*, and MM* with free energies of association $\Delta G(0)$, $\Delta G(1)$, and $[\Delta G(0) + \Delta G(1)]/2$, respectively.

In the model that assumes a homogeneous population at all degrees of dissociation [5] it seems appropriate to postulate a linear dependence of the loss of free energy on degree of dissociation:

$$\Delta G(\alpha) = \alpha[\Delta G(1) - \Delta G(0)] = \alpha\delta G \qquad \text{(3a)}$$

or

$$K(\alpha) = K(0)\exp(\alpha\delta G/RT) \qquad \text{(3b)}$$

We have then as equivalent to Eq. (1)

$$\frac{4\alpha^2 C}{1 - \alpha} = K(0)\exp\left(\frac{\alpha\delta G}{RT}\right) \qquad \text{(4)}$$

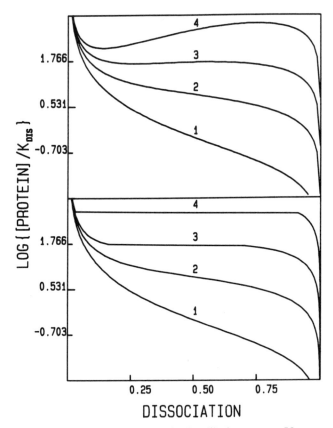

Figure 2. Critical phenomena in the dilution curves. Upper panel: close form procedure of calculation; lower panel: iterative calculation. 1 to 4 correspond to $\delta G/RT = 0, 3, 7, 10$. From ref. [10].

where $K(0)$ is the dissociation constant at negligible dissociation. Taking α as the independent variable and $\delta G/RT$ as parameter the dilution curves calculated by Eq. (4) have the form shown in Figure 2. As dG/RT increases the logarithmic span decreases and at $\delta G/RT = 3 + \sqrt{8} = 5.8284$ the plot includes a vertical segment at which $d \log C/d\alpha = 0$. For larger values of $\delta G/RT$ the plots fold as in the van der Waals isotherms below the critical point, a characteristic seen in critical phenomena, like a condensation of a vapor, when analyzed by a closed-form mathematical procedure. As we have shown elsewhere [10] the dependence of the degree of dissociation on dilution, given by Eq. (3a), can also be calculated by an iteration procedure that employs Eqs. (1) and (3b), the result of

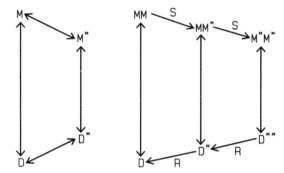

Figure 3. Classical (left) and time-dependent (right) dimer–monomer equilibria involving two dimer and two monomer forms. The differences arise from the postulation of the absence of detailed balance in the conversions of M into M* and D into D*.

which is shown in the lower panel of Figure 2. If $\delta G/RT$ is less than the critical value of 5.83 this procedure gives results identical to those of Eq. **(4)**. For larger values of $\delta G/RT$ there is no "van der Waals fold" but a discontinuous jump in α with $\log C$ indicating that in these cases the dissociation has a segment of abrupt transition similar to that observed in genuine first-order phase transitions. The relatively large value of dG/RT required for the appearance of an abrupt transition makes unlikely the observation of this phenomenon in simple dimer–monomer or tetramer–monomer equilibria. However, the formulated iteration procedure may find an appropriate place in the description of the cooperative association of monomers into proteins made up of many subunits such as apoferritin, or the virus capsids.

In the alternative description of the dimer dissociation with drift the existence is postulated of three fixed values of the free energy of association of well-defined dimer species [11]. At protein concentration such that $\alpha \rightarrow 0$ the predominant forms of dimer and monomer will be MM and M, respectively, and with increasing dilution the predominant dimers will be of the types M*M, first, and finally M*M*. The ordinary way of stating the equilibrium relations when conformationally different forms are postulated is shown in the left scheme of Figure 3. The classical assumption with regard to forms such as M and M* that *can be* interconverted by first-order processes is that, at equilibrium, they exist in fixed proportions determined by a unique isomerization constant $k^0 = [M^*]/[M]$ independent of α. If this assumption is accepted then the dilution curve has a fixed logarithmic span of 2.86 units. It simply represents the dissociation on dilution of a protein in which the species D and M are in respective equilibrium with isomers D* and M*.

Within the classical formulation of the equilibrium there is simply no way of representing the possible transformation through dissociation and reassociation of a protein in which D and M are the predominant forms at high concentration and D* and M* the forms present at low concentrations. However, a study of the reversible dissociation of oligomeric proteins by hydrostatic pressure and the reversible cold inactivation of enzymes (Chapter XV) shows that these transformations actually take place. The difficulty inherent in the classical formulation is, as stated above, that M and M* must exist in a fixed ratio independent of α. Such assumption does not seem justified when applied to the dissociation of a dimer in which the conformations of the monomers in the dimer are determined by the precise contacts that the two make with each other. Upon dissociation the effective cause of many of the details of the monomer conformation within the dimer disappear and the monomer must adopt others that are characteristic of its isolation from the companion subunit. Recovery of the original conformation can take place only when, on reassociation, the characteristic subunit contacts of the dimer are reestablished in a time-dependent process. Therefore it is not reasonable to postulate that the isolated monomer shuttles back and forth between its original form in the dimer, M and other forms, although the information necessary to adopt the M form no longer exists, and cannot be recovered until association takes place. Similarly D and D* cannot be considered as two species in equilibrium present in constant ratio. The stable associated form is D and after the drifted monomers associate they will form a different dimer (D* = MM* or D** = M*M*) that will eventually evolve to regenerate D. A description of the dependence of the observed degree of dissociation on the concentration of protein must therefore include the time required for the transformations M $\rightarrow$ M*, D*D and D** $\rightarrow$ D. If the times for these transformations are comparable, and slow in relation to the time for a cycle of association and dissociation [10], there will be a very stable dependence of the free energy of association on the concentration (see below). However, without appeal to the absolute time dependence of the different forms, the dilution curve can be calculated on the assumption of the existence of fast equilibria between dimers and monomers of each separate type, and of unidirectional transformations M $\rightarrow$ M*, at the spoil rate S and D* $\rightarrow$ D and D** $\rightarrow$ D* at the rate of recovery R (right side scheme of Fig. 3). The fast independent equilibria of the dimer forms with their respective monomers are, in this case described by the relations:

$$K = \frac{4\alpha_0^2 Cf_0}{1 - \alpha_0}; \qquad K_1 = \frac{4\alpha_1^2 Cf_1}{1 - \alpha_1}; \qquad K_2 = \frac{4\alpha_2^2 Cf_2}{1 - \alpha_2} \qquad (5)$$

In Eq. (5) the αs and Ks are, respectively, the degree of dissociation and the dissociation constant of the molecular species with free energies of association $\Delta G(0)$, $[DG(0) + (DG(1)]/2$, and $\Delta G(1)$. The fs are the fractions of the total protein in the three forms. The rates of spoil $(M \rightarrow M^+)$ and repair $(D^* \rightarrow D)$ are in principle independent quantities, but the observable departure from the standard monomer–dimer equilibrium is important only when the corresponding rates of spoil and repair are of comparable magnitude, and is a maximum when these rates are equal. A simple iterative procedure, which actually imitates the kinetics of the system, may be used to solve for the fractions of the three possible dimer forms, f_0, f_1, and f_2, respectively, and the corresponding degrees of dissociation α_0, α_1, and α_2 [9]. With any arbitrary initial set of f values, Eqs. (5) are solved to give α_0, α_1, and α_2. The changes in fs resulting from the transfer between the independent monomer–dimer equilibria are

$$df_{01} = (S\alpha_0 f_0/2) - R\alpha_1 f_1; \qquad df_{12} = (S\alpha_1 f_1/2) - R\alpha_2 f_2 \qquad (6)$$

The new f values are determined by the relations

$$f_0 = f_0 + df_{01}; \qquad f_1 = f_1 + df_{12} - df_{01}; \qquad f_2 = f_2 - df_{12} \qquad (7)$$

The iteration is continued until $df < d\epsilon$ where $d\epsilon$ is a preestablished small value that satisfies the conditions

$$d\epsilon \ll S; \qquad d\epsilon \ll R \qquad (8)$$

At this point the system is in equilibrium within an error $d\epsilon$, and the apparent degree of dissociation is

$$\langle \alpha \rangle = \alpha_0 f_0 + \alpha_1 f_1 + \alpha_2 f_2 \qquad (9)$$

This procedure can be adapted to the description of equilibria involving aggregates of higher order, such as those of tetramers and the corresponding monomers. The dilution curves obtained from Eqs. (5) to (9) differ from those obtained under the assumption of a population with continuously variable free energy of association: the logarithmic span decreases systematically as $\delta G/RT$ increases, but it is always smaller than for the

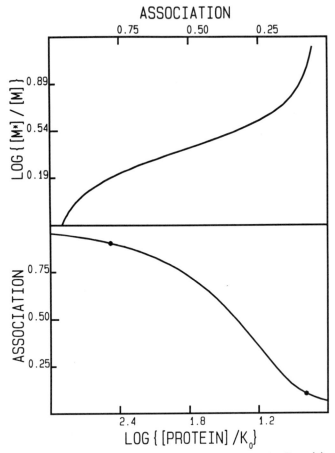

Figure 4. Lower panel: Dilution curve calculated by Eqs. (5) to (9) with $\Delta G(1) - \Delta G(0) = 4.6RT$, The shortened concentration span (1.7 log units) is indicated by the two points on the curve. Upper panel: Dependence of the ratio M*/M for the same case.

homogeneous case (lower panel of Fig. 4). Additionally there is no indication of any critical behavior. These differences between the two models are particularly interesting in view of the fact that in both cases there is a linear change in the apparent free energy of association with the extent of reaction. The variation of the ratio [M*]/[M] as a function of $\langle \alpha \rangle$ is shown in the upper panel of Figure 4. The shape of this plot is independent of the magnitude of the free energy loss as a result of drift and is similar for dimers, trimers, or tetramers. Figure 4 shows that the ratio of the monomers in the two forms increases rapidly only in the range of dissociations greater than 0.5, and very rapidly for $\langle \alpha \rangle > 0.8$. As a consequence we expect that demonstration of the existence of a drifted form through the changed properties of the monomers will require observations at degrees of dissociation that exceed one-half.

One-Dimensional Stochastic Simulation of the Classical Equilibrium and the Equilibrium with Drift

The procedures just described to calculate the dilution curves under the assumptions of the loss of free energy of association as the degree of dissociation increases assume, but do not demonstrate, that an actual equilibrium is reached under these conditions. To carry out this demonstration it is necessary to simulate the equilibrium in a form in which the time dependence of the different molecular species is made explicit, and for this purpose we can carry out a simplified stochastic simulation, which we presently describe. Simulation of a concentration-dependent association requires a three-dimensional random walk of the monomers within the volume of the solution. To simplify the simulation so that it can be carried out in a reasonable time in a desktop computer it is convenient to restrict the monomer random walk to a single dimension [12]. We do not expect to lose a great deal by this substitution: The time-dependent characteristics that the process of random diffusion confers to the monomer association may be expected to be similar, at least in qualitative fashion, in one and three dimensions. Additionally, as the diffusion is isotropic within the volume, the one-dimensional diffusion can be held to correspond to a process of radial diffusion in space. In the one-dimensional simulation the space allowed for diffusion consists of a line of L cells that are numbered 1 to L. If 0 and 1 are randomly drawn numbers, 1 may be held to correspond to motion to the next cell with higher number and 0 to motion to the next lower numbered cell. When a monomer occupies the first or Lth cell the next step is always a reflection regardless of the random number drawn. The monomers are independently moved after a random number is drawn for each of them, the motion is executed, and formation of the dimer is deemed to occur when both monomers share the same cell number. The dimer is assumed to have an average lifetime equivalent to N steps. A random number between 1 and N is drawn and one of these numbers, say 1, determines the dissociation, while drawing the other numbers result in the persistence of the dimer for one more step. When the dimer dissociates, as a result of drawing 1, two further random numbers between 1 and L are drawn to position the newly born monomers in the diffusion space and a new AD cycle begins. In each AD cycle the numbers of steps that the dimer and monomers have survived are counted. After a sufficiently large number of AD cycles the degree of dissociation is calculated by the equation

$$\alpha = \frac{M}{M + 2D} \tag{10}$$

where M is the total number of steps performed as monomer and D is the total number of steps as dimer. The number of steps L allowed for monomer diffusion bears a direct relation to the concentration C employed in an actual experiment: The cube of the average distance between the molecules is proportional to the volume in which they are enclosed and therefore inversely proportional to C. We can then write

$$C^* = \frac{1}{L^3} \tag{11}$$

where C^* differs from C, the actual concentration, by a simple proportionality constant. Therefore if our one-dimensional simulation offers an acceptable representation of the classical dimer–monomer equilibrium with a dissociation constant independent of the degree of dissociation, a plot of $\log(1/L^3)$ against the degree of dissociation should match in all important respects the dilution curve predicted by Eq. (1). The lower panel of Figure 5 shows the dilution curve calculated from Eq. (1), and the points resulting from the one-dimensional stochastic simulation. The logarithmic span is 2.89 units.

It seems indispensable to inquire more closely to the reasons for the rather remarkable success of the representation of the three-dimensional diffusion by the simple one-dimensional model just described. We expect a dependence of the number of average random steps required for association on the square of the length of the one-dimensional box. The upper panel of Figure 5 shows such dependence: The slope of the plot of M against L^2 is 0.135, so that to a good approximation we can set

$$M = \left(\frac{L}{e}\right)^2 \tag{12}$$

The appearance of e in the denominator of this expression is evidently related to the existence of a Poisson distribution of initial distances between the diffusing monomers. With the number of monomer steps given by (12) the degree of dissociation of the dimer placed in a box of

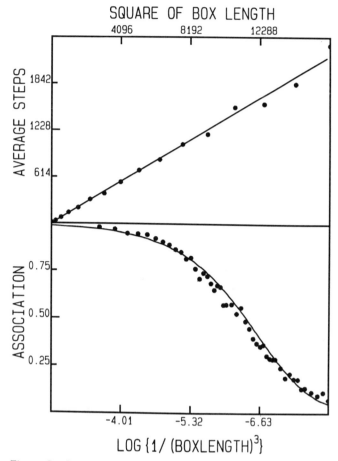

Figure 5. Lower panel: Stochastic simulation of the equilibrium without drift. Upper panel: Dependence of
the average number of steps before monomer
association on the square of box length.

length L is

$$\alpha = \frac{L^2}{L^2 + 2De^2} \qquad (13)$$

This simple expression for α is useful in that it allows one to reliably
anticipate many characteristics of the equilibrium without having to per-

form the lengthy stochastic simulations. With α given by the last equation and C replaced by $1/L^3$ Eq. (1) becomes

$$K' = \frac{4L}{D'(D' + L^2)} \tag{14}$$

where we have set $D' = 2De^2$. The average dissociation constant $\langle K \rangle$ over the whole dilution curve is

$$\int_{\alpha=0}^{\alpha=1} K'(\alpha)\, d\alpha = \int_{L \ll D'}^{L \gg D'} \left[\frac{8L^2}{(L^2 + D')^3} \right] dL \tag{15}$$

The last expression for $\langle K \rangle$ is integrated by parts giving

$$\langle K \rangle = \frac{\pi}{2D'^{3/2}} \tag{16}$$

and the ratio

$$\frac{K'(\alpha)}{\langle K \rangle} = (2/\pi)[(\alpha(1-\alpha))]^{1/2} \tag{17}$$

is independent of D'. Equation (17) shows that K' is not a constant but a very slowly varying function of α, symmetric about $\alpha = 1/2$. K'/K equals 1.27 at $\alpha = 1/2$, becomes unity at $\alpha = 1/2 \pm 0.3098$, and reaches 0.556 at $\alpha = 1/2 \pm 0.45$. Thus over the interval in degrees of dissociation that is experimentally meaningful (~ 0.05–0.95) Eq. (16) may be expected to give results that agree with thermodynamic prediction well within the experimental uncertainties of the typical case. We note also that the span of $\log(1/L^3)$ between symmetric values of α is identical to that predicted by the thermodynamic formulation, namely $\log\{[\alpha/(1-\alpha)]^3\}$.

A simulation of the association–dissociation equilibrium *with drift* demands the specification of the rates of the processes of spoil and repair. The treatment is simplified by the introduction of conditions derived from the experimental observations, the most important of which comes from estimations of the relative rates of dissociation and association of dimer proteins. The dissociation observed after either sudden dilution or application of pressure takes hours to several minutes to be accomplished while the reassociation after release of pressure is fast in comparison with the time taken to perform the measurements. Further, it appears that, in dimers at least (Chapter XV), the increase in dissociation constant on application of pressure largely follows the increase in the rate of dissociation. We may try then to simulate the process under the simplifying assumption that the rate of association corresponds to a diffusion-controlled process, whereas the dissociation rate reflects directly the affinity of the monomers for each other. We expect that the rate of dissociation of a dimer immediately after its formation depends on the extent of "spoiling" that took place when the monomers were separated, and that the spoiling will be repaired and the unperturbed dissociation rate recovered, partially or totally, during the lifetime of the dimer. The monomers generated in the dissociation can then inherit some spoiling from the dimer and it may require a number of AD cycles before a stationary state between the rates of spoiling and repair is reached. It is the inherited nature of the processes of spoil and repair that necessitates a stochastic simulation. To pass from the case of a unique dissociation constant to the one in which it is dependent on the degree of dissociation we need to specify the maximum extent of spoiling as well as the rates of spoiling during the monomer lifetime, and repair during the dimer lifetime. The maximum extent of spoiling, or correspondingly the minimum lifetime of the dimer, may be gauged from the increase of the dissociation constant necessary to explain the shortened logarithmic spans experimentally observed, or the decrease in affinity that follows the reassociation of the oligomers following their dissociation by application of moderate pressures (Chapter XV). The observed changes of the dissociation constant after decompression are of one to two and one-half orders of magnitude corresponding, under our assumptions, to a shortening of the dimer lifetime by a factor of 10 to several hundreds. In the simulations we have taken this as the typical case. For some simple arguments [5], as well as from the stochastic simulations themselves, it follows that the larger changes in apparent dissociation constant require that the rates of spoil and repair be of comparable magnitude. Accordingly we shall assume that they are equal so that a specified change in the dimer lifetime requires the same number N of steps of monomer lifetime (spoil) or dimer lifetime (repair).

The results of the simulations as plots of degree of dissociation against $\log(1/L^3)$ are displayed in Figure 6. If the rates of spoil and repair are made exceedingly fast so that these processes are completed in the first step after dissociation (spoil) or association (repair), that is if we set $N = 1$, there is an insignificant shortening of the logarithmic span. When the number of steps necessary for spoiling and repair becomes larger the span of the dilution curve decreases as expected from Eqs. (3)–(10).

The analysis that we have made of the dynamic character of the association of monomers stands in contrast to the belief, implicit in most of the literature on oligomeric proteins, that we can expect to understand the varied phenomena of protein association on the simple assumption of rigidly defined oligomeric structures that interact in a predefined way to

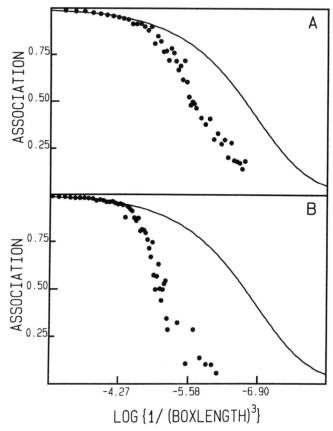

Figure 6. Stochastic simulation of dimer–monomer equilibria with conformational drift. (A) Free energy loss = $2.52RT$. (B) Free energy loss = $4.96RT$.

yield the specific aggregates. If the analysis of the experimental data given above is correct we must always consider the time-dependent character of the associations. Particularly in the case of aggregates of many particles the time-dependent effects may be dominant of the whole process and confer to it the character of an abrupt transition.

Time-Dependent Equilibria and the Principle of Detailed Balance

Since the classical publications of Onsager of 1931 [14] it has been customary to accept as generally valid that in the state of thermodynamic equilibrium "every microscopic process and its reciprocal occur with equal frequency." Onsager recognized that detailed balance cannot be derived from thermodynamic constraints and that it must be introduced as an independent assumption. In his publication he referred only to first-order processes of interconversion of chemical forms and did not explicitly consider the possibilities that would arise in chemical equilibria in which both first-order and second-order equilibria are interlocked, so that one becomes the cause of the other. We note that according to the formulation given above we can interpret the principle of detailed balance in two different ways: In the more lax formulation it simply states that we can expect in an AD cycle the reversal of all the microscopic processes that maintain it, without reference to the point of the cycle at which any such reversal occurs. Such formulation is not contested by the observations of the dependence of the free energy of association on the extent of reaction. In the more stringent formulation of the principle of detailed balance the processes of conformational change that accompany the dissociation and association would be constrained to occur simultaneously with these processes: It forsees that if we could record a moving picture of the microscopic events taking place at equilibrium we should not be able to distinguish whether the film is being run forward or backward. The conformational changes characteristic of the isolated monomer would appear in complete synchrony with the process of dissociation and the recovery of the characteristics of the dimer would similarly accompany the association. In consequence we would be unable to tell from it which are the causes and which are the effects at the molecular level. If we believe that the association of the monomers and their separation are the *specific causes* of the conformational changes that follow, we are obliged to accept the common sense view that both the independence of the chemical potentials from the extent of reaction and the more stringent formulations of the principle of detailed balance correspond to limiting conditions that apply only when the rates of change of molecular conformation are fast in comparison with the rates of association and dissociation. To quote Einstein [15]: "No fairer destiny could be allotted to any physical theory than it should of itself point the way to a more comprehensive theory in which it lives on as a particular case."

References

1. Breslow, E. (1979) *Annu. Rev. Biochem.* **48**, 251–274.
2. Rholam, M., and Nicolas, P. (1981) *Biochemistry* **20**, 5837–5843.
3. Scarlata, S., and Royer, C.A. (1986) *Biochemistry* **25**, 4925–4929.
4. Brewer, J. (1981) *CRC Crit. Rev. Biochem.* **11**, 209–254.

5. Xu, G.-J, and Weber, G. (1981) *Proc. Natl. Acad. Sci. U.S.A.* **79**, 5268–5271.

6. Shore, J.D., and Chakrabarthy, S.K. (1976) *Biochemistry* **15**, 875–879.

7. Silva, J.L., Silveira, C.F., Correia, A., and Pontes, L. (1992) *J. Mol. Biol.* **223**, 545–555. See also Bowie, J.U., and Sauer, R.T. (1989) *Biochemistry* **28**, 7139–7143.

8. Triose phosphate isomerase: Waley, S.G. (1973) *Biochem. J.* **135**, 165–172; β_2-tryptophan synthase: Blond, S., and Goldberg, M.E. (1985) *J. Mol. Biol.* **182**, 597–606; $\alpha\alpha$-tropomyosin: Mo, J., Holtzer, M.E., and Holtzer, A. (1991) *Proc. Natl. Acad. Sci. U.S.A.* **88**, 916–920.

9. Pyruvate carboxylase: Irias, J.J., Olmsted, M.R., and Utter, M. (1969) *Biochemistry* **8**, 5136–5148.

10. Weber, G. (1987) *Proc. Natl. Acad. Sci. U.S.A.* **84**, 7359–7362

11. Weber, G. (1986) *Biochemistry* **25**, 3626–3631. For a different approach to the same problem see Cooper, A. (1988) *J. Mol. Liquids* **39**, 195–206.

12. Weber, G. (1989) *J. Mol. Liquids* **42**, 255–268.

13. Silva, J.L., Miles, E.W., and Weber, G. (1986) *Biochemistry* **25**, 5780–5786.

14. Onsager, L. (1931) *Phys. Rev.* **37**, 405–413. Onsager, L. (1931) *Phys. Rev.* **38**, 2265–2279.

15. Einstein, A. (1961) *Relativity*. Engl. Trans. R.W. Lawson. Crown, New York, p. 77.

XV

The Effect of Pressure on Oligomeric Proteins and Protein Aggregates

Josephs and Harrington [1] observed the apparent decrease in the sedimentation constant of myosin solutions as the rotor speed of the ultracentrifuge was increased and deduced that the excess hydrostatic pressure associated with the spinning of the sample caused the dissociation of myosin aggregates. Since then similar sedimentation behavior has been demonstrated in a number of oligomeric proteins. The pressures generated in the analytical ultracentrifuges are of the order of 100 atmospheres and therefore the effects are conspicuous only in those cases in which the standard volume decrease on dissociation is sufficiently large [2]. Considerably higher pressures may be reached in preparative, swinging bucket high-speed centrifuges and more recently chemical and enzymic changes in proteins have been observed following sedimentation and attributed to the pressure-induced dissociation of oligomers [3].

Phenomenological Description of the Pressure Dissociation

Consider again the dissociation of the simplest aggregate, a protein dimer. At a pressure p the dissociation constant K_P is related to the degree of dissociation α, and the protein concentration as dimer, C by the equation

$$K_P = \frac{4\alpha^2 C}{1 - \alpha} \tag{1}$$

K_P is related in turn to the dissociation constant at atmospheric pressure and to ΔV, the standard volume change on association of the monomers, by the relation

$$K_p = K_{atm} \exp\left(\frac{p\Delta V}{RT}\right) \tag{2}$$

Equation (1) may be put in the form

$$-\log C = f(\alpha) - \log K \tag{3}$$

with $f(\alpha) = \log[4\alpha^2/(1 - \alpha)]$, and employing Eqs. (1) and (2)

$$\frac{p\Delta V}{2.302\,RT} = f(\alpha) - \log\left(\frac{K_{atm}}{C}\right) \tag{4}$$

Thus the dependence of the dissociation on the logarithm of the concentration at constant pressure, and the dependence on the pressure, at constant concentration have equivalent forms. Corresponding to the characteristic logarithmic span of the concentration we have a characteristic pressure span, $\langle dp \rangle$, the difference between pressures at which 10 and 90% dissociation are achieved. From Eq. (4)

$$\langle dp \rangle = \frac{2.302\,RT \log(729)}{\Delta V} \tag{5}$$

where we have assumed, as we shall generally do in what follows, that ΔV is independent of the applied pressure. This assumption is reasonable, in view both of the very limited compressibilities of proteins [4] and the available experimental data on their pressure dissociation. From Eqs. (1) to (5) it is evident that by observations on the pressure effects of oligomeric proteins we can extract two important parameters: The change in volume upon association, ΔV, and the dissociation constant at atmospheric pressure, K_{atm}. ΔV is determined from the slope of the plot of $f(\alpha)$ against p [Eq. (4)] or simply from the pressure span $\langle dp \rangle$ [Eq. (5)]. We note that agreement between the two methods implies that ΔV is either pressure independent or that the change in ΔV with pressure is not large enough to produce a deviation from linearity in the plot according to Eq. (4) [5]. Linear extrapolation to $p = 1$ bar, according to Eq. (4), can be used to determine K_{atm}. We recall that the dilutions necessary to obtain appreciable dissociation at atmospheric pressure are so large that the most obvious method for the determination of the dissociation constant, the change in association on dilution, cannot in general be applied. On the other hand, if we take $\Delta V = 150$ ml mol^{-1} and $p = 2.3$ kbar Eq. (2) gives $K_P/K_{atm} = 1.1 \times 10^6$. For a typical value of $K_{atm} = 10^{-9}$, a micromolar solution has, by Eqs. (1) and (2), a degree of dissociation 0.016 at atmospheric pressure and 0.996 at 2.3 kbar. Thus, on application of a pressure at which we expect minimal effects on the conformations of the separated monomers, we can pass from almost complete association to virtually complete dissociation and we can estimate with some precision the value of K_{atm}.

Demonstration of the Dissociation: Measurement of ΔV and K_{atm}

Detection of the dissociation of protein aggregates under pressure has been achieved by other methods besides those involving fluorescence: Light scattering was used by Payens and Heremans to demonstrate the disaggregation of the casein particles [6]. Paladini *et al.* [7] have constructed an instrument to carry out electrophoresis at pressures under 3 kbar and have used it to demonstrate the dissociation of β_2-tryptophan synthase, and also RNA polymerase. By observations of preparations fixed under pressure with glutaraldehyde Silva and Weber [8] demonstrated the reversible dissociation of the capsids of Brome mosaic virus and other viruses. The measurement of enzyme activity after release of pressure has been used as an indication of dissociation during the preceding compression by Jaenicke and co-workers [9], but this method can be relied on only when the aggregates are totally active and the subunits completely inactive, irrespective of time and magnitude of the previously applied pressure. When degrees of dissociation after decompression are measured by an independent method and compared with the residual enzyme activity it is clear that most of the cases so far studied do not fall in this category.

Determination of the degree of dissociation under seeming equilibrium conditions has been to now limited to optical spectroscopic, mainly fluorescence, observations. These can involve measurements of the changes in the center of mass of the emission, and of the polarization of the fluorescence, both from the fluorophores intrinsic to the protein or from extrinsic ones covalently attached to it [2]. Whatever method we employ we must ensure that the changes in the spectral properties arise from actual dissociation of the particles and not from a condition independent of it, as could be a first-order reaction of the protein brought about by the pressure that changed fluorescence yield, spectral distribution, or polarization. The distinction is easily made: A dissociation reaction must depend on the concentration in a predictable way [2], whereas a first-order reaction will be strictly independent of the protein concentration. Demonstration of the concentration dependence of the spectral property is conclusive proof of the dissociation, and is regularly observed in dimers and tetramers. If two protein concentrations C_1 and C_2 of a dimer are employed, the difference in pressures $dp(C_1, C_2)$ at which the same degree of dissociation is reached is, from Eq. (4)

$$dp(C_1, C_2) = \left(\frac{2.302\,RT}{\Delta V}\right) \log\left(\frac{C_1}{C_2}\right) \qquad (6)$$

If ΔV is pressure independent a concentration increase simply shifts the curve toward higher pressures without changing its shape. Equation (**1**) may be generalized to an aggregate of n particles in which the only species present in appreciable amounts are the aggregate and the monomers, to give

$$K_p = \frac{n^n \alpha^n C^{n-1}}{1 - \alpha} \qquad (1')$$

and correspondingly

$$\frac{p \Delta V}{2.302\, RT} = f_n(\alpha) - \log\left(\frac{K_p}{C^{n-1}}\right) \qquad (4')$$

with $f_n(\alpha) = \log[n^n \alpha^n /(1 - \alpha)]$. Corresponding to Eq. (**6**) we have

$$dp_n(C_1, C_2) = (n - 1)\left(\frac{2.302\, RT}{\Delta V}\right) \log\left(\frac{C_1}{C_2}\right) \qquad (6')$$

Equation (**6'**) is obeyed by dimers whereas tetramers show less pressure displacement by a change in concentration than this equation predicts, and aggregates of many subunits often show a minimal or even virtually absent concentration dependence. The origin of these important differences is discussed in detail below.

The covalent conjugates of proteins with fluorophores of appropriately long lifetime, by the use of which one can determine directly the average rotational relaxation time of the particles [10], occupy a special place in the demonstration of the dissociation as they can give a measure of the apparent volume of the particles by application of Perrin's law (Chapter XII). In these cases there is no need for the observation of plateau polarizations at atmospheric and high pressure, but one requires instead the value of the limiting polarization of the fluorophore, and of the fluorescence lifetime. The demonstration of dissociation by this method has been done in the dimers of yeast enolase, tryptophan synthase of *Escherichia coli* and yeast hexokinase and in the tetramers of lactate dehydrogenase, yeast glyceraldehyde dehydrogenase, and phosphorylase A. In each of these cases, except in phosphorylase A, the dissociation could also be inferred by the changes in spectral shift of the fluorescence. In all these diverse cases changes in the polarization of the intrinsic protein fluorescence were observed. Silva *et al.* [11] showed that the dissociation of the β_2 subunit of tryptophan synthase can be followed equally well by measurement of the spectral shift of the tryptophan fluorescence, by changes in the fluorescence of the natural prosthetic group, pyridoxal phosphate, or by the polarization of the fluorescence of dansyl covalent conjugates. In phosphorylase A the thermodynamic parameters of the tetramer–monomer dissociation found by the polarization of intrinsic fluorescence, the polarization of dansyl and pyrene butiryl conjugates, and the transfer of electronic energy between subunits labeled with fluorescein residues (see below) were in good agreement [12].

Equation (6), that permits calculation of the volume change on association from the displacement of the plots along the pressure axis as the total protein concentration is changed is obeyed—within an error in ΔV of 10–20%—by a number of dimer molecules. The lack of more precise agreement with theory is not clear; it may depend on a small contribution from compressibility, or, more likely, from the approximate nature of the expressions that relate, in the different methods, the degree of dissociation to the observed spectral parameters. According to Eqs. (6) and (6′) we expect that, for equal values of ΔV, the parallel displacement of the pressure dissociation curve of a tetramer with concentration will be three times larger than for a dimer. In the tetramers this far studied, lactate dehydrogenase, yeast glyceraldehyde-phosphate dehydrogenase, and phosphorylase A, the parallel displacements are *3 to 5 times smaller* [13, 14] than those predicted by Eq. (6′). In other words there is, for the tetramers, a large discrepancy between the values of ΔV calculated from Eqs. (4′) and (6′), whereas for the dimers there is reasonable agreement between them. Computation shows that contributions from neither the increased compressibility of the monomers nor from conformational drift can quantitatively account for the discrepancy, but it is possible to account for it on the assumption that the tetramers form a heterogeneous population as regards the free energy of association [14]. We expect that in this case the plot of the dissociation against the applied pressure results from the superposition of several curves with different values of K_{atm} and the same value of ΔV (Fig. 1) and the real ΔV could be several times larger than the value computed from the logarithmic span.

Causes of the Pressure Dissociation of Oligomers

The dissociation of protein aggregates subjected to pressure has been found to be a very general phenomenon shown by the indefinite aggregates of myosin, actin, and tubulin [15], the capsids of icosahedral viruses [8], and a number of simple dimers and tetramers. It is then expected that the causes of this behavior are to be found in very general features of the structure of proteins. This impression is strengthened by the reduced range of values of ΔV and K_{atm} so far encountered in the dissociation of dimers and tetramers: The data of references [2] and [11]–[14] show that ΔV is comprised in the range of 50–300 ml mol^{-1} and K_{atm} in the range

HETEROGENEOUS TETRAMER

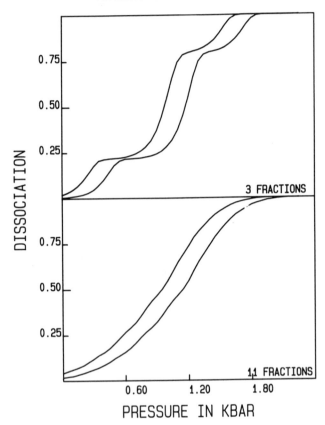

Figure 1. Effect of the heterogeneity of a tetramer population on ΔV_P and ΔV_C. The plots are for two
protein concentrations differing by a factor of
10. Upper plots: Three fractions in proportions
1:3:1. Free energies of association are $+9RT$, 0,
$-6RT$. Lower plots: Eleven fractions. Skewed
Gaussian distribution with standard deviations
$+7.5RT$ and $-5RT$. These latter plots give
$\Delta V_\mathrm{P} = 208$ ml mol^{-1}, $\Delta V_\mathrm{C} = 800$ ml mol^{-1}.

of 10^{-7} to 10^{-11}. Paladini and Weber [2] proposed as one cause of the
pressure-induced dissociation the existence of voids or dead spaces between the subunits that would disappear when the neighboring subunit is
replaced by solvent, as a result of the better molecular packing in the
latter case. Noticeably this cause can apply equally well to the dissociation
of oligomeric proteins under pressure and to the denaturation of proteins

made up of a single peptide chain. X-Ray crystallography has identified in globular proteins cavities of varied size [16] and Lewis and Rees [17] note that the intersubunit surfaces in concanavalin A and superoxide dismutase are significantly more uneven than the surfaces of these proteins that contact the solvent.

A second influence, and probably a very important one, arises from the existence of salt linkages at the intersubunit boundaries. They clearly appear in the X-ray crystallographic structures of hemoglobin and lactate dehydrogenase [18] among other oligomeric proteins, but the importance of electrostatic linkages in holding the subunits together can also be surmised from the dissociation of many protein aggregates at extremes of pH or at high ionic strength. On dissociation the charges at the newly formed subunit surfaces undergo hydration with a consequent decrease in volume of the system as a result of electrostriction.

A third cause of decrease in volume on subunit dissociation should follow from the replacement of interactions between nonpolar groups at the two sides of the intersubunit boundary by interaction of each of these with the permanent water dipoles. However, as Bøje and Hvidt [19] have pointed out, the decrease in volume of apolar molecules on dissolution in water, though particularly noticeable in very dilute solutions, can markedly decrease and even invert with increase in concentration, and the density of the nonpolar groups that are exposed on dissociation in oligomeric proteins is such that they may be best represented by the latter cases. Also, the "interior" of the protein contains a large number of peptide dipoles and cannot be considered as an apolar medium [20]. Replacement of water for deuterium oxide leads systematically to a decrease in the value of K_{atm} with no appreciable change in ΔV, an additional pointer toward the limited importance of the cause last mentioned.

In conclusion, although all the causes that we can easily enumerate favor dissociation of the aggregates under pressure we are for the moment unable to decide on their relative importance, both in general and in most of the particular cases. The difficulties in assigning specific changes in molecular interaction energies to the decrease in volume on dissociation may best be appraised by noticing that in a dimer of molecular weight 100,000 we expect the area of contact between the monomers to be ~ 1600 Å^2, and a change in volume of 150 ml/mol results in a change in volume of 250 Å^3 per molecule. Thus, on dissociation the atomic contacts between the subunits are replaced by others that, on average, shorten interatomic distances by 0.15 Å.

The specific substitution of amino acid residues at the boundaries of oligomeric proteins carried out by the methods of genetic engineering should yield, at least in the favorable cases, the more precise data that are needed to decide on the relative contributions of the boundary interactions to the change in volume and in free energy on association. Observations by Sligar and co-workers [21] on the effect of amino acid substitution on the volume change in the formation of the specific complex between cytochrome b_5 and cytochrome c clearly show that in this case the electrostriction of hydrated ion pairs plays a major role in the determination of the volume change on dissociation. As shown in Figure 2 substitution of a charged residue by an uncharged one produces a decrease in association volume that is of the same order (25 ml mol^{-1}) as that determined for the dissociation of acetic acid in water [22].

Kinetics of Pressure Dissociation

Few data exist at present of the rates of association and dissociation of the particles. By following the changes in fluorescence polarization of tryptophan, Xu and Weber [23] found that on dilution at atmospheric pressure the dissociation equilibrium of enolase is completed in the course of hours. King and Weber [24] found that after a rapid raise in pressure to 2 kbar complete dissociation of bovine or porcine lactate dehydrogenases was reached in about 30 min and Ruan and Weber have found that half dissociation of glyceraldehyde-phosphate dehydrogenase [25] is reached some 6 min after the rise in pressure. The increase in dissociation constant at elevated pressures may result from an increase in the rate of dissociation k_-, a decrease in the rate of association k_+, or both these causes. We can then set

$$k_-(p) = k_- \exp\left(\frac{fp\Delta V}{RT}\right); \qquad k_+(p) = k_+ \exp\left[\frac{(f-1)p\Delta V}{RT}\right] \quad (7)$$

where k_- and k_+ are the corresponding rates at atmospheric pressure and f a fraction between 1 and 0 that *formally* determines the partition of the effects of pressure on the two rates: If $f = 1$ the increase in $K(p)$ is wholly caused by the increase in the rate of dissociation and if $f = 0$ it depends on the decrease in the rate of association alone. f can be determined by a study of the relaxation to equilibrium after rapid successive pressure increases [26]. It is difficult to conceive that f is other than constant, or a very slowly varying function of the pressure, and a practical estimation of f must start with the assumption that its value is independent of pressure, in which case the relaxation after two successive pressure increases will suffice for its determination. For ease and accuracy of operation these two increases should be chosen to have maximum amplitude and to that purpose the pressure must be rapidly raised from atmospheric to that of mid-dissociation, $p_{1/2}$, and after equilibrium at this

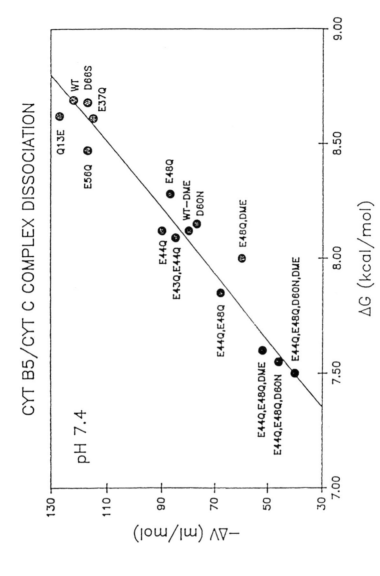

Figure 2. Correlated changes in ΔV and ΔG following substitutions of the interacting charged amino acids in the complex of cytochrome b_5 with cytochrome c. From ref. [39].

pressure is achieved, raised once more to the pressure of nearly complete dissociation. The corresponding relaxations are characterized by the relaxation times t_1 and t_2 at which e^{-1} of the corresponding amplitudes of the changes in dissociation are reached. The effect of an "instantaneous" increase in pressure is calculated as follows: The differential change $d\alpha$ in time dt is given, from Eq. (1), by

$$d\alpha = \left\{ k_-(p)(1 - \alpha)C - 4k_+(p)\alpha^2 C^2 \right\} dt \qquad (8)$$

where $k_-(p)$ and $k_+(p)$ correspond to the rates of dissociation and association, respectively, at the new pressure p. Introducing the values of the rates of association and dissociation given in Eq. (7), (8) takes the form

$$d\alpha = \left\{ 1 - \alpha - 4\alpha^2 \left(\frac{C}{K_0} \right) \exp\left(\frac{p_{1/2}\Delta V}{RT} \right) \right\} \exp\left(\frac{fp_{1/2}\Delta V}{RT} \right) k_- C \, dt \qquad (9)$$

Starting with a value of C/K_0 appropriate to yield a small dissociation α_0 at atmospheric pressure, the increment $d\alpha$ over a suitably small time interval dt (time units $= 1/\{Ck_- \exp(fp_{1/2}\Delta V/RT)\}$ is calculated by the last equation and added to a_0 to obtain a new value of the dissociation, a_1. The procedure is repeated until middissociation is asymptotically approached, and the time for e^{-1} change in dissociation is interpolated in the set of values of α versus time. The whole procedure described by Eq. (9) is carried out once more starting with the equilibrium value of α and the last equation is applied after substituting $p_{1/2}$ by the pressure at which nearly complete dissociation is calculated to occur. As shown in Figure 3, $t_1/t_2 = 0.68$, if $f = 0.1$ and 28.0 if $f = 0.9$, a large difference that permits a ready practical estimation of the relative effects of pressure on the association and dissociation rates. A similar calculation can be carried out for a tetramer that dissociates into monomers through an equilibrium in which the dimer concentration is negligible. Equation (8) is still valid for this case if α^2 is replaced by α^4 and C refers to the protein concentration as tetramer. It is then found that $t_1/t_2 = 0.61$, if $f = 0.1$ and 75.4 if $f = 0.9$. For both dimers and tetramers $\log(t_1/t_2)$ has a strict linear dependence on f, and Figure 4 shows the values of f determined from this dependence in the dimers and tetramers studied. It appears that in virtually all cases the effects of pressure are on both the rates of association *and* dissociation, but to very different extents in dimers and tetramers: in dimers pressure mainly accelerates the dissociation, whereas in the tetramers pressure *appears* to slow down the rate of association preferentially. However, we note that by reason of the assumptions made in the application of Eq. (9) to tetramers the rate of association k_+ refers

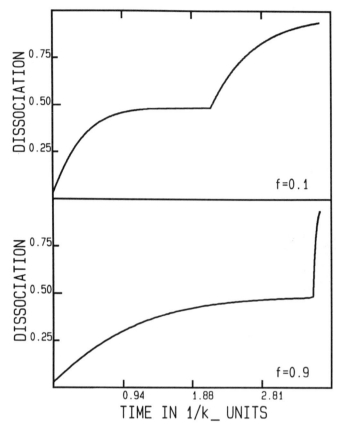

Figure 3. Differences in the relaxation to equilibrium depending on the preferential effect of the pressure on the dissociation rate ($f = 0.9$) or the association rate ($f = 0.1$) in a dimer, calculated by Eq. **(9)**.

to the whole process of formation of the four-subunit particle from the monomers and includes therefore the rate of dissociation of the intermediates formed in the course of tetramer association. Thus, although the results of Figure 4 show a genuine difference in the effects of pressure in the respective cases of dimers and tetramers they do not in any way exclude the possibility that the main effect of application of pressure is, in both cases, that of accelerating the various rates of dissociation of the overall particle and intermediate aggregates. In dimers and tetramers decompression is followed by reassociation within the dead time of the measurements (1–2 min) indicating an approximate rate of association of 10^6 mol^{-1} sec^{-1}. The observed rates of association and dissociation of the

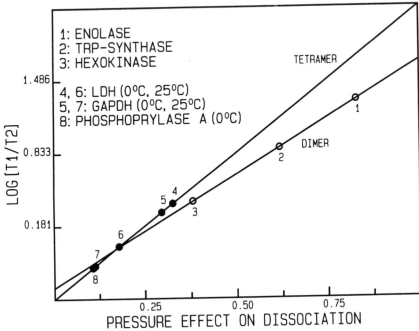

Figure 4. Apparent relative changes in the rates of association and dissociation by the applied pressure. f, of Figure 3, was determined by placing the experimental $\log(t_1/t_2)$ on the theoretical straight lines for the dimers and tetramers. Numerical key: 1: enolase, 2: β_2-tryptophan synthase, 3: hexokinase, 4,5: GAPDH, 6,7: porcine LDH, 8: glycogen phosphorylase. 1, 3, 5, 7, and 8 at 0°C, 2, 4, and 6 at 25°C [26].

β_2 dimer of tryptophan synthase are consistent with the thermodynamic dissociation constant at atmospheric pressure, $K_{atm} = \sim 10^{-9}$ determined from the equilibrium dissociation data [11]. "Immediate" reassociation on decompression has also been seen in bovine and porcine lactate dehydrogenases, glyceraldehyde-phosphate dehydrogenase, and phosphorylase.

Kinetics of Subunit Exchange in Oligomeric Aggregates

It is possible to obtain information on the dynamics of the cyclic process of association and dissociation of oligomeric proteins by following the exchange of subunits among the aggregates at various degrees of dissociation, including those small enough to be undetected by the methods described above. A general method applicable to protein aggregates made up of both identical and different subunits relies on the transfer of electronic energy between appropriate fluorophores covalently attached to the protein [26]. Two different procedures have been used: In one, transfer takes place between two different fluorophores with appearance of sensitized fluorescence (heterotransfer method); in the other depolarization of the fluorescence follows transfer of the excited state among identical fluorophores (homotransfer method). In the heterotransfer method identical solutions of the protein are covalently labeled with the donor and the acceptor fluorophore and the conjugates are purified by conventional methods. Immediately after mixing the solutions the fluorescence is excited at a wavelength of preferential absorption by the donor and the fluorescence spectrum is recorded. As the subunit exchange requires a cycle of dissociation and association that usually takes a much longer time than the 1 or 2 min needed to acquire the fluorescence spectrum, this zero time spectrum can be considered characteristic of the absence of transfer. After a variable time after mixing, under the conditions of concentration, temperature, and pressure chosen for the experiment, the spectra are recorded and the amount of transfer is quantified by resolving the spectrum into its components. In the homotransfer method the protein solution is labeled with fluorescein isothiocyanate (or another fluorophore with similar spectroscopic characteristics) to the extent that, ideally, each subunit carries a single label. Transfer of electronic energy between the identical fluorophores attached to the protein results in depolarization of the emitted radiation [26]. Energy transfer between identical fluorophores fails when excitation is carried to the red edge of the absorption [27] and this circumstance provides a useful check of the origin of the depolarizing effects. After mixing with an appropriate excess of unlabeled protein the polarization increases as a result of segregation of labeled subunits into different particles as new oligomers are formed by reassociation of labeled and unlabeled subunits. Employed in this manner energy transfer methods permit the direct determination of τ_x, the average time for subunit exchange, a measure of the stability and permanency of the aggregates under specified conditions of temperature, pressure, and solvent composition. As described below they also permit a direct comparison of the rates of dissociation and subunit exchange under pressure, and therefore they allow a decision to be made as to the character, stochastic or deterministic, of the equilibrium of the aggregate and its subunits.

Modification of the Dissociation by Added Ligands: Energetic Coupling between Ligand and Subunit Associations

The changes in the free energy of association of the subunits on the addition of specific ligands have only begun to be explored. Ruan and Weber [14] followed the pressure dissociation of yeast hexokinase in the presence of the substrates and products of the catalyzed reaction, and of

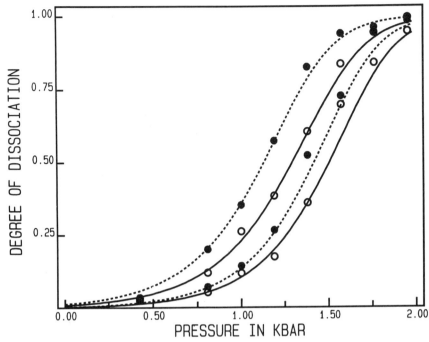

Figure 5. The effects of physiological ligands at saturating concentrations (10^{-2} M) on α vs. pressure in hexokinase [14]. From left to right the theoretical dissociation curves are those for ATP: $\Delta G = -11.95$, $\Delta V = 163$; no additions: $\Delta G = -12.4$, $\Delta V = 156$; glucose: $\Delta G = -13.5$, $\Delta V = 174$; glucose + AMPPCP: $\Delta G = -13.75$, $\Delta V = 170$. ΔG values in kcal mol^{-1}, ΔV values in ml mol^{-1}.

the pseudo substrate AMPPCP, at saturating concentrations and the results are presented in Figure 5. The nearly parallel dissociation curves show that ΔV is not significantly modified, while their appreciable displacements indicate directly the stabilization, or otherwise, of the monomer association by the added ligand. The changes in the free energy of association at atmospheric pressure varied from $+0.4$ kcal mol^{-1} (ATP) to -1.35 kcal mol^{-1} (glucose + AMPPCP). Their absolute values are very much what we expect from previous estimates of interligand free energy couplings (Chapters III, VIII, and IX). Evidently pressure dissociation provides a direct, and potentially accurate, method for the determination of the energetic couplings between subunit boundary and binding site; therefore it holds promise of providing some pointers as to their physiological significance in each particular case. In yeast hexokinase we note two important features: One is that the coupling free energies are on the order of 1 to 3 RT units; the other is that the energetic effects of the ligands on

Table 1. Effects of Ligands at Saturating Concentrations
on the Free Energy of Association at Atmospheric
Pressure, ΔG_{atm}, and Association Volume of Yeast
Glyceraldehyde-Phosphate Dehydrogenase[a] [25].

Ligand	ΔG_{atm} (kcal mol^{-1})	ΔV (ml mol^{-1})
None	−34.9	230
ATP	−31.2	237
ADP	−30.8	178
Cyclic AMP	−36.2	241
NAD	−37.2	241
Glyceraldehyde 3-phosphate	−35.1	224

[a]From ref. [25].

the free energy of association, of the monomers, $\delta\Delta G$(ligand), are
not additive: thus $\delta\Delta G$(glucose) = −1.15 kcal mol^{-1}, $\delta\Delta G$(ATP) =
$\delta\Delta G$(AMPPCP) = +0.4 kcal mol^{-1}, and $\delta\Delta G$(glucose + AMPPCP) =
−1.35 kcal mol^{-1}. Thus glucose and ATP produce contrary boundary
effects, but when both are present the stabilization is increased over that
for glucose alone. This result appears in reasonable agreement with the
investigations on the enzyme kinetics, according to which the binding of
glucose needs to precede that of ATP to result in the catalytically
productive binding of the latter ligand [28]. Similar observations on the
modifications of the pressure dissociation by the addition of ligands have
been carried out with the tetramer of yeast glyceraldehyde-phosphate
dehydrogenase [25] and these results are shown in Table 1.

Apart from the remarkable differences observed, for example, between
$\delta\Delta G$(ATP) or $\delta\Delta G$(ADP) and $\delta\Delta G$(c-AMP), we note that the larger free
energy couplings in the tetramer are three to four times greater than in
the yeast hexokinase dimer. This feature may be expected on grounds that
the effects of the ligands are exerted on four intersubunit surfaces rather
than on one as is the case with the dimer. We expect that the multiple free
energy couplings play a role similar to that encountered in the case of the
chemiosmotic conversion (Chapter IX).

Hysteresis of the Dissociation and Conformational Drift
of the Separated Subunits

Perhaps the most interesting finding to emerge from all the studies
made to now of the dissociation of oligomeric proteins under pressure is

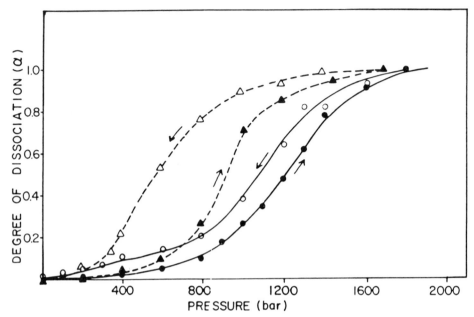

Figure 6. β_2-Tryptophan synthase: Hysteresis increases and subunit affinity decreases at 5°C (– – –) compared to 25°C (——). Full circles and triangles are equilibrium dissociations at increasing pressures; open symbols for decreasing pressure [11].

that separation of the protein subunits is followed, in virtually every case, by conspicuous conformation changes. This conclusion can be reached from at least three different kinds of observations:

1. The curve giving the dependence of the degree of dissociation with increasing applied pressure differs from the curve obtained on decreasing the pressure from that at which complete dissociation is reached. The ascending pressure branch is, in most of the cases studied, systematically displaced to lower pressures with respect to the descending pressure branch indicating that the reassociated species have decreased subunit affinity. This characteristic hysteretic behavior is seen clearly in the dissociation of the β_2 subunit of tryptophan synthase (Fig. 6), in porcine or bovine lactate dehydrogenases, in yeast hexokinase, and in yeast, rabbit muscle, and *E. coli* glyceraldehyde-phosphate dehydrogenases. Yeast GAPDH shows the strongest effects and these are shown in Figure 7.

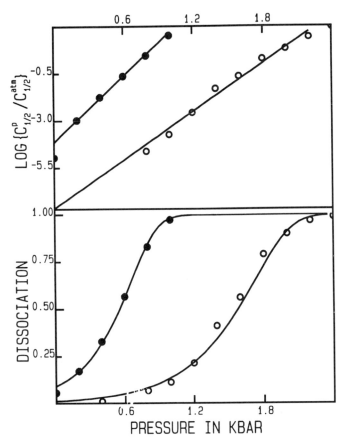

Figure 7. Hysteresis effects on GAPDH. Open circles: increasing pressure. Filled circles: decreasing pressure. Lower panel: dissociation plots. Upper panel: plots of relative free energy changes. The experimental data [25] are fitted by curves with the parameters $C/C_{1/2} = 180$, $C/C^*_{1/2} = 12$, $\Delta V = 235$, $\Delta V^* = 320$ ml mol^{-1}.

2. On rapid decompression from the pressure of complete dissociation to atmospheric pressure complete reassociation of subunits takes place "immediately" in some cases: in other cases the immediate reassociation is only partial. However, in both cases the recovered reassociated protein shows very reduced enzyme activity. Both association and enzyme activity increase with time and reach, in most cases, the value before compression. Such an

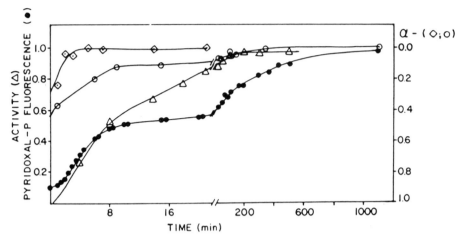

Figure 8. Recovery of original properties after decompression in β_2-tryptophan synthase. Speed of recovery decreases in the order fluorescence polarization (rhombs), fluorescence spectrum (open circles), enzyme activity (triangles), pyridoxal fluorescence (filled circles) [11].

effect is particularly evident in the lactate dehydrogenases where recovery of activity at room temperature may take many hours [9, 24]. The time for recovery of the enzyme activity differs widely depending on the protein and temperature. At room temperature yeast glyceraldehyde-phosphate dehydrogenase recovery is complete in about 1–3 hr [25] and in the β_2 subunit of tryptophan synthase recovery has a half time of about 10 min [11].

3. Other properties that depend on protein conformation such as the fluorescence spectrum of tryptophan or the fluorescence yield of pyridoxal phosphate in β_2-tryptophan synthase show slow recovery on decompression, but over times periods that differ conspicuously from both the reassociation of the subunits and the regaining of enzyme activity (Fig. 8).

From these observations we may conclude that on dissociation of the aggregate the subunits undergo a progressive change in conformation losing partially the affinity for each other and that on reassociation a conformational adjustment that restores the original properties of the aggregate takes place. From the fact that the recovery of different properties takes place following different time courses one is forced to conclude that the separated subunits exist in a variety of conformations rather than in a unique one, thus undergoing after disaggregation a process of "conformational drift."

Numerical Simulation of the Hysteresis Effects

There is little reason to doubt that the equilibria established at the different pressures involve average conformations, and therefore average chemical potentials, of the aggregate and free subunits that vary with the degree of dissociation. Before attempting a quantitative description of the equilibrium that makes use of this notion it is convenient to look into the dependence of the mean life of dimers and monomers on the degree of dissociation in an *ordinary* dimer–monomer equilibrium governed by Eqs. (1)–(4) and therefore not subjected to conformational drift. The change of the mean life of the dimer or monomer at equilibrium under pressure is determined by the fractional effects of the pressure increase on the rate constants of association and dissociation, described by the factor f of Eqs. (7)–(9). Taking as a specific example an equilibrium with the characteristic pressure-dissociation parameters (ΔG_{atm} and ΔV) of β_2-tryptophan synthase, if the pressure exerts its effects through the increase in the rate of dissociation exclusively ($f = 1$), then the dimer lifetime drops by a factor of 10^6 at the pressure of nearly complete dissociation and the monomer life *also decreases*, by a factor of 40. When $f = 0.5$ and the pressure effects are equally exerted on the dissociation and association processes the lifetime of the dimer decreases by a factor of 140 and the lifetime of the monomer now *increases* by a factor of 35. If $f = 0$ the dimer lifetime remains constant and that of the dimer increases by a factor of 4×10^4. The pressure dependence of the half-lives of monomer and dimer under pressure is illustrated in the panels of Figure 9. It is evident that if $f = 1$ the conformational drift will be determined exclusively by the decreased lifetime of the dimer, therefore by the time allowed for the repair of the drift, which will take place after association. As f decreases, the ensuing lengthening of the monomer life permits a larger degree of spoiling and in the limit of $f = 0$ we expect that even the unchanged lifetime of the dimer becomes insufficient for the repair process. We note that because of these differences we expect the drift process to be substantially different in the dissociation by dilution and in the dissociation by pressure. In the absence of drift, if dissociation is brought about by dilution, the lifetime of the monomer increases as 1/concentration and the lifetime of the dimer remains constant, whereas in the pressure dissociation the lifetimes of both monomer and dimer change in opposite directions even if f is as large as 0.75. Also we may expect differences between dimers and tetramers as regards their respective conformational

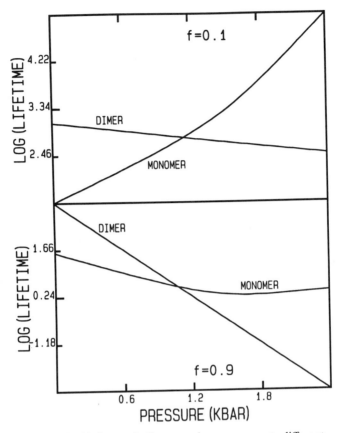

Figure 9. Half lives of dimer and monomer at different pressures, when the pressure effects are preferential on the dissociation ($f = 0.9$) and association ($f = 0.1$). Constant free energy of association [Eqs. (4) and (9)] has been assumed.

drifts because, grossly, $f > 0.5$ for the former and $f < 0.5$ for the latter (Fig. 4).

Evidently the method of choice to describe such time-dependent equilibria is one of stochastic simulation, already described in Chapter XII, but a simpler phenomenological treatment that is able to reproduce the principal features of the phenomena of hysteresis and drift makes use of the coupled equilibria of aggregates formed by the normal (M) and drifted (M*) monomers, already discussed in Chapter XIV, and described by Eqs. (XIV.5) to (XIV.9). To simplify the simulations we eliminate the intermediate MM* and restrict the dimers to only two types, D and D*. Equilibration at any fixed pressure involves the iteration procedure described by Eq. (**XIV.6**) but with a single df that must be made smaller than $d\epsilon$. The dissociation constants of the species D and D* at pressure p are given by the expressions:

$$\text{For D} \longleftrightarrow 2\text{M}, \qquad K(p) = K_{\text{atm}} \exp\left(\frac{p\Delta V}{RT}\right) \qquad (10)$$

$$\text{For D*} \longleftrightarrow 2\text{M*}, \qquad K^*(p) = K_{\text{atm}}^* \exp\left(\frac{p\Delta V^*}{RT}\right) \qquad (11)$$

The values of the degrees of dissociation at a pressure p_J, α_J, and α_J^*, respectively, are used as initial values in the calculation of the dissociations α_{J+1} and α_{J+1}^*, as a result of pressure p_{J+1}, and the procedure starts with α_{atm}, the degree of dissociation at atmospheric pressure. If the pressures are first increased to obtain complete dissociation and then decreased to atmospheric pressure, always satisfying Eq. (**XIV.6**) with the conditions $d\epsilon \ll S$ and $d\epsilon \ll R$, the pressure dissociation curves for ascending and descending pressure are identical [29]. This condition will in general necessitate N iterations and results in unique values of the components at each pressure, whether this has been reached by increase from atmospheric pressure or decrease from the pressure of virtually complete dissociation. If the number of iterations is now decreased below N so that the condition $S > d\epsilon > R$ is satisfied, the curve corresponding to ascending pressures progressively differs from that corresponding to descending pressures, as experimentally observed. The dissociations at descending pressures reflect the persistence of the forms D*, M* the only ones present at complete dissociation under the assumption $S > d\epsilon > R$. By choice of appropriate values of the ratios C/K and K/K^*, ΔV and ΔV^* different effects of the persistence of the drifted forms are obtained. The decompression profile is determined by both the ratios K^*/K and $\Delta V^*/\Delta V$, and Figure 10 shows two cases that have found experimental representation in the oligomeric proteins so far studied. In both cases the free energy of association of the drifted form is decreased ($K^*/K = 100$). In the case that is apparently the more common, ΔV^* exceeds ΔV (LDH, GAPDH, β_2-tryptophan synthase) but cases in which $\Delta V^* < \Delta V$ have also been observed (hexokinase, triose-phosphate isomerase).

Of particular interest is that the complete recovery of the aggregation at atmospheric pressure is strongly dependent on both C/K and K^*/K. This difference is dramatically demonstrated by a comparison of the holo- and apoproteins of β_2-tryptophan synthase at the same (micromolar) concentration: The former ($K^*/K \simeq 10$) reassociates completely but only one-half of the latter ($K^*/K > 100$) does so immediately after decompression [11].

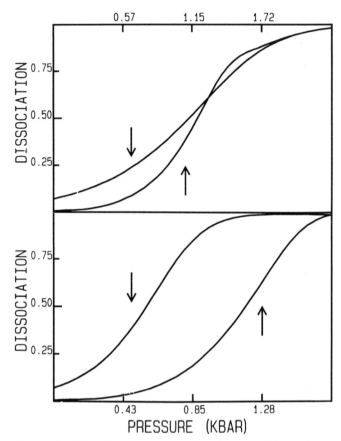

Figure 10. Showing the importance of the association volume of the drifted form in the character of the hysteresis of the pressure-dissociation cycle. The simulations employ Eqs. **(10)** and **(11)** and the parameters $\Delta V = 150$ ml mol^{-1}, $K:K^*:K^{**} = 1:10:100$, $S = 0.02$, $R = 0.0001$, 150 iterations. Upper plot: $\Delta V^{**} = 200$ ml mol^{-1}. Lower plot: $\Delta V^{**} = 100$ ml mol^{-1}.

Character of the Dissociation Equilibria

The dissociation into subunits requires diffusion of the particles away from each other and the state favorable to dissociation must therefore have a lifetime sufficiently long to permit such diffusion: The minimum distance, dx, at which we can consider that the particles resulting from the dissociation become independent of each other will be when separated by a layer of water molecules, that is about 5 Å units. From Einstein diffusion equation

$$\langle dx^2 \rangle = 2D\,dt \tag{12}$$

A diffusion coefficient D of 10^{-6} cm^2 sec^{-1}, which corresponds to a particle of some 35 kDa in water at 25°C, gives $dt > 1$ nsec. Thus the "dissociating state" that permits the particles to diffuse away from each other must have a lifetime longer than 1 nsec. Clearly this state cannot be one of transient accumulation of energy in some degree of freedom, as it is now clear that such a state, typically a vibrational state of some sort, would not last beyond some picoseconds [30]. What is required is a transient fluctuation of structure endowed with an appropriate lifetime. A reasonable candidate [29] is a fluctuation of charge that produces a greatly diminished interaction energy at the boundary or even an actual electrostatic repulsion. Fluctuations of charge, affecting the carboxyl groups at the intersubunit boundary, must take place and it is known that the half-life of a protonated carboxyl group is a few microseconds, therefore quite sufficient to permit separation of the particles by diffusion.

Figure 11 depicts a possible scheme for the various steps of the equilibrium between a dimer carrying a fluctuating quadrupole disposition of charges at the boundary and the constitutive monomers. Since the dissociation is brought about by a relatively rare fluctuation of boundary charges that results in repulsion there is no reason to expect that reassociation will occur only when the same form is adopted; this form actually *opposes* monomer association, which we further know to occur at a rate not much lower than the rate of monomer encounters. At low degrees of dissociation the limited lifetime of the monomers (Fig. 9) may not result in any appreciable change in average structure of the dimer and the equilibrium is limited to the forms in the upper half of the diagram. At larger degrees of dissociation the decreased lifetime of the dimers, the increased lifetime of the monomers, or both permits a significant degree of conformational drift to persist after reassociation and yields the modified dimer shown in the lower half of the diagram. Upper and lower halves of the diagram convey the concept that we are dealing with protein forms to which different chemical potentials must be attributed according to the degree of dissociation at which the equilibrium becomes established. This conclusion is at variance with the proposition, current since the original formulation by Gibbs, that the chemical potentials are constants, without reference to the particular extent of reaction at which the chemical equilibrium is established. Notice that a similar conclusion as regards the variability of the chemical potential has been reached in the study of the

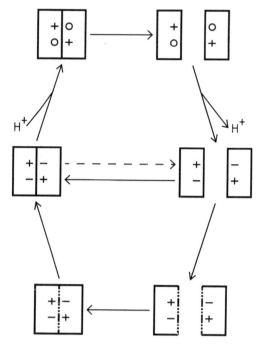

Figure 11. Dissociation of a dimer carrying a
fluctuating quadrupole at the interface between the subunits, discussed in text.

dissociation by dilution, at atmospheric pressure, of two dimers as discussed in Chapter XIV. The variation of the chemical potential of the
protein particles with the degree of dissociation is further strengthened by
other observations apart from those already mentioned.

1. In comparing the pressure dependencies of the polarization of
    dansylated covalent adducts of lactate dehydrogenase and the
    spectral distribution of the intrinsic protein fluorescence King and
    Weber [24] noticed that ΔV equals ~ 170 ml mol^{-1} in the
    former case and close to twice this value in the latter. The
    polarization values of the dansyl conjugates reflect primarily the
    average volume of the observed particles and may be expected
    to correspond more closely to the degree of dissociation. The
    observed difference in ΔV indicates that the changes in confor-
    mation responsible for the changes in the environment of the
    tryptophan do not exactly correspond to the appearance of the

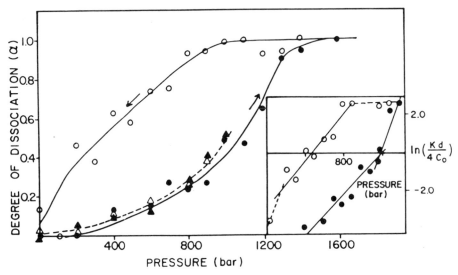

Figure 12. Asymmetry of the pressure dependence of the dissociation, and dependence of the hysteresis of the apoprotein of β_2-tryptophan synthase [11] on the degree of dissociation reached under pressure. Circles: pressure decreased after complete dissociation. Triangles: pressure decreased after reaching 50% dissociation. Filled symbols: ascending pressure. Open symbols: descending pressures.

dissociated form; instead they take place over a restricted range of degrees of dissociation and lead to an incorrectly high value of ΔV.

2. Silva *et al.* [11] have shown that both holo- and apoprotein of the β_2 dimer of the tryptophan synthase from *E. coli*, have similar values of ΔV, 160 ml mol^{-1}, but that the apoprotein has not only a diminished free energy of association but exhibits a very asymmetric dependence of the degree of dissociation on pressure. Whereas 900 bars are required to increase the degree of dissociation from 0.1 to 0.5 less than a quarter of this value is sufficient to pass from 0.5 to 0.9 (Fig. 12). As indicated in Figure 12, the hysteresis observed on complete dissociation of the apoprotein is very marked, whereas it is virtually absent if the maximum dissociation reached is 0.5. This difference in behavior corresponds to that predicted by the upper and lower halves of Figure 11, and by the dependence of the relative lifetime of dimers and monomers on the degree of dissociation of Figure 10. In contradistinction

the dependence of $\log f(\alpha)$ on p for the holoprotein shows a minimal asymmetry.

All the considerations that we have enumerated in this section show the unexpected complexities that appear in the simplest of time-dependent equilibria and explain the reluctance of experimenters and theoreticians to consider them, and their preference for the idyllic world of time-independent opposing processes.

The Combined Effects of Temperature and Pressure

In both the lactate dehydrogenases and in β_2-tryptophan synthase decreasing the temperature results in a decrease in subunit affinity and more evident hysteresis. The ratio of the dissociation constants at 5° and 25°C, at atmospheric pressure, reaches 60 for lactate dehydrogenase [24] and about 10 for β_2-tryptophan synthase [11]. Even larger effects are observed in yeast hexokinase [14] and in glyceraldehyde-phosphate dehydrogenase [25]. Figure 13 shows plots of two complete cycles of compression and decompression of a solution of GAPDH carried out at 0°C with intervals of 2 hr: at 0°C in one case and at 20°C in the other. The "drifted form" appears to be permanently stable at 0°C, but reverts almost completely into the species not previously subjected to pressure on standing for 2 hr at 20°C. Similar drift and recovery are observed with lactate dehydrogenase [24]. The enzymic activity of both proteins subjected to the pressure cycle was very small but both recovered when the temperature was raised to 25°C.

The dissociation constants at atmospheric pressure, K_{atm}, of GAPDH and phosphorylase A obtained by extrapolation of the plots of $f(\alpha)$ against pressure are shown in Table 2. They indicate that the stability of the tetramer decreases markedly with temperature, and so far this characteristic has been found to be generally applicable to protein aggregates of many kinds.

Table 1 shows that although the free energies of association at 25°C are nearly equal the decrease in subunit affinity with temperature is much more marked in phosphorylase A. In a vant'Hoff plot of the data the slopes of $\Delta G/T$ against $1/T$ have opposite signs and correspond to $\Delta H = -11.4$ kcal mol^{-1} for GAPDH and $\Delta H = +21.2$ kcal mol^{-1} for phosphorylase A. The corresponding entropy contributions, $T\Delta S$, at 25°C are $+21.2$ and $+53.7$ kcal mol^{-1}, respectively. A positive enthalpy of association is seen in hexokinase and a negative one in β_2-tryptophan synthase, but in all cases the association of the monomers into specific aggregates is driven by the considerable entropy increase in the reaction. Even in yeast glyceraldehide-phosphate dehydrogenase, if the entropy

GAPDH: TWO SUCCESSIVE COMPRESSION CYCLES

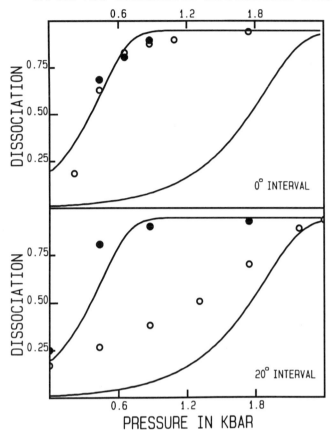

Figure 13. Effect of a second cycles of compression and decompression of GAPDH at 0°, with an interval of 2 hr after the first cycle [25]. Upper panel: Interval at 0°C. Lower panel: Interval at 20°C. Open circles: ascending pressures. Filled circles: descending pressures.

contribution were negligible the protein would be dissociated at all biologically relevant concentrations, in spite of the considerable negative enthalpy change.

What is the molecular origin of this large and apparently universal entropy of association? The general opinion has been in the past (see Ben-Naim, Chapter XII, [1]) that on dissociation the interaction of the apolar amino acid residues at the new surfaces of contact with the water dipoles generates a large entropy decrease of the water, by stabilizing its dipole association. However, there is little direct experimental evidence

Table 2. Free Energies of Dissociation of Yeast Glyceraldehyde-Phosphate Dehydrogenase[a] and Glycogen Phosphorylase A[b] at Several Temperatures

Temperature (°C)	GAPDH (kcal mol^{-1})	Phosphorylase A (kcal mol^{-1})
−5	−30.3	
0	−31.9	
2		−27.9
5	−32.4	
7		−29.6
8	−32.6	
14		−30.7
15	−33.3	
16		−30.9
20		−31.7
25	−33.7	−33.5
30	−34.1	−33.6

[a]From ref. [25].
[b]K.-C. Ruan (unpublished).

that this is the case. Because of its considerable variation from one protein
to another it seems more plausible that much of the excess entropy of the
aggregate is to be sought in the increased mobilities of the amino acid
residues when the largely apolar subunit interfaces are in contact with
each other, with respect to their mobility when they contact the water
dipoles. This is a possibility first suggested by Sturtevant [31]. An investigation of the Debye–Waller factors of the atoms at the subunit interfaces of
crystals of oligomeric proteins should greatly contribute to clarifying this
problem.

Conformational Drift and the Cold Inactivation of Oligomeric Enzymes

At atmospheric pressure micromolar solutions of lactate dehydrogenase
or glyceraldehyde-phosphate dehydrogenase lose most of the enzymic
activity after a few weeks at cold-room temperature (4°C) but regain it
completely after a few hours at room temperature (25°C). The degree of
dissociation remains very small, virtually undetectable, during the cold
inactivation period and the subsequent room temperature reactivation.
However, both processes are concentration dependent: the rate of cold
inactivation decreasing with concentration and the rate of room-tempera-
ture reactivation increasing with concentration [32]. Cryoinactivation has
been observed with many oligomeric proteins that are tetramers and

aggregates of higher order, whereas no monomeric or dimeric enzyme has been reported to be cold sensitive.

The decreased stability of lactate dehydrogenase, at atmospheric pressure, at cold-room temperature in comparison with room temperature has been demonstrated by means of hybridization experiments of the H_4 and M_4 isozymes. Starting with the intact enzyme solutions at room temperature decrease in temperature and increase of pressure produce similar hybridizing effects. King and Weber [32] proposed that both pressure inactivation and cryoinactivation are brought about by a process of dissociation, conformational drift of the separated subunits, and reassociation into inactive aggregates similar to those observed after high pressure dissociation followed by decompression. At low temperature the inactive aggregates predominate because of the expected high energy of activation required to convert the drifted aggregates into the original native species; at room temperature reactivation is rapid enough to maintain the enzymic activity and the higher subunit affinity characteristic of the original aggregate. A free energy scheme that describes this formulation is shown in Figure 14. It follows our view, discussed in detail in Chapter XIV, that the equilibria of the aggregate and its subunits are best described by independent equilibria of the native and drifted species, linked by rates of spoil in the monomer and repair in the dimer. Figure 14 shows the Gibbs free energy of the aggregate and monomer as a function of a conformational coordinate. If we start with tetramers having a unique conformation, the free monomers with which the aggregate is in equilibrium will undergo relaxation toward their stable form M^*. At various stages of the relaxation, dependent on their mean free life, the monomers will reassociate to form aggregates that will evolve toward the distribution corresponding to the minimum free energy. The equilibrium thus established will inevitably comprise heterogeneous populations of both monomer and aggregate. For a sufficiently long monomer free life, which implies a large degree of dissociation, the monomers will reach the equilibrium conformation M^* and the aggregates A^* formed by monomers in this condition will represent the bulk of the aggregates present. Whereas the rate of spoil may be assumed to be independent of the number of subunits in the aggregate the rate of repair will be critically determined by the complexity of the relations of each subunit with its neighbors in the aggregate. It is thus understandable that the persistence of drifted aggregates in tetramers is much greater than those in dimers and that the drift process may appear as practically irreversible in the larger and complex aggregates of the virus capsids. The existence of a broad population of conformations in the native, enzymically active initial population of tetramers appears then as a natural consequence of the differences in conformation of the free and bound monomers rather than as an *ad hoc* assumption to explain the discrepancy in the values of the change in volume of association measured

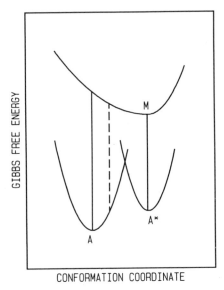

Figure 14. Gibbs free energy graph of the equilibrium between aggregate and subunits showing the energy relations of original (A) and drifted forms (A*) of the aggregate responsible for the conformational drift.

by the pressure span and by the concentration dependence of the pressure dissociation profile (Fig. 1).

Because of the stability of the drifted form at low temperature, one may expect that these modified aggregates will be eventually crystallized, thus permitting the determination of their structure by X-ray analysis. Indeed the crystallization of proteins is often achieved at low temperature and high salt concentration, conditions that must favor the incipient dissociation of aggregates and their reassociation in the drifted, enzymically inactive form. The small difference in the free energy of association of the native and drifted forms could be easily compensated by relatively weak crystal forces making the drifted form permanently stable in the crystal lattice. It is then not at all unlikely that some of the crystallographic tetramer structures already determined belong with the drifted rather than the active species. Liddington and collaborators [33] have recently shown that oxygenated hemoglobin may be crystallized in a structural form that shows a wholly different disposition of the subunits and many differences at the atomic level in comparison with the one originally determined by Perutz and his collaborators. A recent observation of Faber and Matthews [34] is that the unit cell of crystals of a mutant T_4 lysozyme contains four molecules with significantly different conformations. It seems reasonable to suppose that the molecular interactions in oligomers, which are bound to be stronger than the intermolecular forces in the crystal, can stabilize aggregates of different conformations.

The very slow cold inactivation of the more concentrated solutions of LDH or GAPDH does not permit one to determine whether the reduced rate of inactivation results from a greatly increased stability of the aggregate or from the effective repair of the incipient drift of the monomers. A study of the subunit exchange by the methods of energy transfer permits a clear choice between these possibilities. If increased stabilization of the tetramer is at work we will expect at high temperature (25°C) an increase in the subunit exchange time, τ_x, with respect to that observed in the solutions at 0°C. The opposite, an increased rate of subunit exchange, can only be consistent with increase in repair since the free energy of association at 25°C is conspicuously larger than at 0°C. Erijman and Weber (unpublished) found that at atmospheric pressure τ_x is some 20 times faster at the higher temperature clearly indicating that a cyclic process is taking place in which the repair in the aggregate fully compensates for the drift in the monomers. This finding agrees with previous observations [11] that the hysteresis and conformational drift effects are much more pronounced at the lower temperatures.

Stochastic and Deterministic Equilibria in the Association of Oligomers

Silva and co-workers [35] have shown that the pressure-induced dissociation of erythrocruorin from the oligochete worm *Glossoscolex paulistus* (MW = 3.1×10^6, 12 octameric subunits or 96 monomer subunits) is entirely concentration independent. Almost complete independence from concentration is observed in the pressure dissociation of the capsid of the Brome mosaic virus [8] and in other icosahedral viruses, and recently in hemocyanin [36]. However, in all these large aggregates apparently stable states of dissociation equilibrium are reached on increasing the pressure over the same range, 0.5–2.5 kbars, that is covered in the dissociation of dimers and tetramers. The conclusion must be drawn that they constitute a heterogeneous population, each member having its own characteristic dissociation pressure, rather than a population of identical polymeric aggregates in dynamic equilibrium with the interacting subunits. The designation of "deterministic equilibrium" [26] for these cases of independence on the concentration of the particles seems justified by the similarity

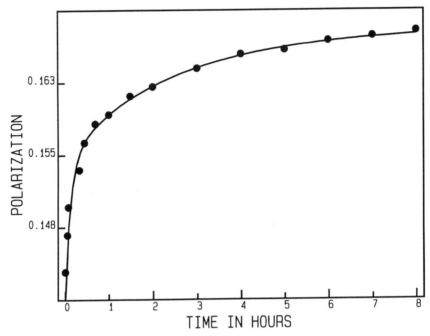

Figure 15. Demonstration of the slow subunit exchange in glyceraldehyde phosphate dehydrogenase by the homotransfer method [26].

with the behavior of macroscopic bodies. In either case the new equilibrium after an external perturbation arises from individual characteristics of each object that are retained over times much longer than those necessary for equilibration. Erijman and Weber [26] examined the dynamics of subunit exchange of dimers and tetramers under conditions of half-dissociation achieved by application of the appropriate hydrostatic pressure. The relaxation time for half dissociation at pressure $p_{1/2}$ [t_1 as in Eq. (7)] is of the order of a few minutes, and in a typical stochastic equilibrium in which the whole population has a uniform probability of dissociation we expect that the time for subunit exchange τ_x would be of the same order. This characteristic behavior is approached in dimers (enolase, hexokinase), but in the tetramers of LDH, GAPDH, and phosphorylase A at low temperature (0° to 2°C) it is found that τ_x is longer than t_1 by more than an order of magnitude. Figure 15 demonstrates this behavior by the time course of the fluorescence polarization of a solution of GAPDH labeled with fluorescein isothiocyanate to the extent of 3 labels per tetramer, at 0°C, in the presence of a 10-fold excess of unlabeled protein, followed over 8 hr. The curve fitting the points is the sum of three exponential terms with relaxation times/amplitudes of 6 min/0.50, 98 min/0.25, and 235 min/0.25. Similar results were obtained with LDH employing the heterotransfer method already described. In phosphorylase A, t_1 was 5.5 min and the subsequent polarization increase, which was seen only when excess unlabeled protein was added, had a characteristic time of 119 min. The observed large values of τ_x/t_1 in the tetramers at low temperatures necessitate two conditions:

1. An appropriately slow rate of monomer reassociation into tetramers.

2. As the 50% pressure-induced dissociation is maintained for many hours, and the remaining tetramers cannot dissociate at a rate any greater than that allowed by the stationary association of the monomers, the existing aggregates must be persistently different from those that dissociated at a much faster rate when the pressure was first raised to that of mid-dissociation.

The existence of fractional differences in pressure stability is further demonstrated by another experiment employing the homotransfer method: Compression to a pressure that results in half-dissociation, followed by decompression and return to 50% dissociation does not result in additional depolarization. It appears that the monomers generated by species of tetramers with a particular free energy of association cannot readily form tetramers by association with monomers that have other free energies of association. In that case the expectation is that at pressure p all the fractions i of the population for which $p\Delta V_i + \Delta G_i \geq 0$ are completely dissociated, whereas those for which $p\Delta V_i + \Delta G_i < 0$ remain unsplit. For further dissociation to occur at pressure p, fluctuations in ΔG must occur that generate new p-sensitive fractions among the undissociated tetramers, and τ_x must reflect the time necessary for the interconversion of these fractions in the population. We can easily envision the conditions that lead to the two extreme cases of stochastic and deterministic equilibrium of the dimers and of the many-subunit particles, respectively. However, it appears difficult to devise a quantitative explanation for the effects observed in the transitional case of the tetramers at low temperature, particularly the extremely sharp pressure discrimination among the members of the population.

Classically, the deterministic character of molecular processes observed on a macroscopic scale is expected to result from the statistics of a sufficiently large number of independent stochastic events, yet there are reasons to expect, or suspect, that at some level in the biological organization determinism at the molecular level replaces stochastic behavior. At present we have no clear idea as to how such deterministic behavior arises or the degree of complexity of the system at which it becomes important. However, deterministic behavior is clearly apparent in the association equilibria of many-subunit aggregates such as virus capsids, erythrocruorin, and hemocyanin and in a reduced, though not less evident form, in the association of monomer subunits to form a tetramer at low temperatures. The complex relations between the parts of a macromolecule appear to lead to a large restriction of the local molecular motions, but only to an extent that permits what in the limit would give individual characteristics to each particle, sufficiently persistent to confer a deterministic character to some of the events in which the macromolecule participates.

Biological Implications of the Conformational Drift

Isolated single chain proteins in solution have been shown to be indefinitely stable and to recover readily the native characteristics when these are lost by moderate heating, by addition of denaturants, and even by reversible disulfide reduction [37]. Following these experiments it has been widely accepted that the amino acid sequence is the determinant of a unique three-dimensional structure, and in the absence of evidence to the contrary this supposition has been assumed to apply to the subunits of oligomeric proteins. The observations discussed in this and the previous chapter show that the latter extension has to be carefully qualified. The properties of oligomers differ according to the previous history of the sample and appear to depend on the concentration, presence of specific ligands, temperature, and pressure over an indefinite time before that at which the properties are measured. Thus oligomeric proteins have to be grouped with other "materials of fading memory" [38] well known in studies of the mechanics of solids. The variety of behavior of such protein materials is exemplified by the hysteresis after decompression, by the anomalous dilution curves, and by the persistence of altered properties at low temperature, all of which result from dissociation of the aggregate followed by drift and reassociation of drifted monomers. It may be appropriate to note that the experiments carried out so far refer to isolated solutions of protein and that conditions may considerably differ in the enormously more complex environment of the cell. As the conformational drift results in progressive loss of catalytic activity and other properties of the oligomers we would expect that special arrangements exist in organisms to prevent, delay, or repair it. Their limited efficiency could furnish an additional reason for the fast turnover of proteins in the organism, surprising when one considers the stability of isolated proteins in solution. The most commonly offered explanation for the fast turnover of proteins is that it is necessary to maintain a protein composition optimal for the changing physiological needs. However, it is also possible that the physiological convenience of rapid replacement is simply an opportunistic by-product of the inherent instability of the oligomeric proteins.

References

1. Josephs, R., and Harrington, W.F. (1967) *Proc. Natl. Acad. Sci. U.S.A.* **58**, 1587–1594.

2. Paladini, A.A., and Weber, G. (1981) *Biochemistry* **20**, 2587–2593.

3. Champeil, P., Buschlen, S., and Guillain, F. (1981) *Biochemistry* **20**, 1520–1524.

4. Gavish, B., Gratton, E., and Hardy, J.C. (1983) *Proc. Natl. Acad. Sci. U.S.A.* **80**, 750–754. Gekko, K., and Hasegawa, Y. (1987) *Biochemistry* **25**, 6563–6571.

5. Bridgman, P.W. (1931) *The Physics of High Pressure*. Dover, New York, pp. 130–135.

6. Payens, T.A.J., and Heremans, H.A.K. (1987) *Biopolymers* **8**, 335–345.

7. Paladini, A.A., Silva, J.L., and Weber, G. (1987) *Anal. Biochem.* **161**, 358–364.

8. Silva, J.L., and Weber, G. (1988) *J. Mol. Biol.* **199**, 149–159.

9. Mueller, K., Luedemann, H.-D., and Jaenicke, R. (1982) *Biophys. Chem.* **16**, 1–7.

10. Weber, G. (1987) In *Current Perspectives in High Pressure Biology*, H.W. Jannasch, R.E. Marquis, and A.M. Zimmerman, eds. Academic Press, Orlando, pp. 235–244.

11. Silva, L.J., Miles, E.W., and Weber, G. (1986) *Biochemistry* **25**, 5781–5786.

12. Ruan, K., and Weber, G. (1991) In press.

13. King, L., and Weber, G. (1986) *Biochemistry* **25**, 3632–3637.

14. Ruan, K., and Weber, G. (1988) *Biochemistry* **27**, 3295–3301.

15. Engelborghs, Y., Heremans, K.A.H., De Mayer, L., and Hoebeke, J. (1976) *Nature (London)* **259**, 686–689.

16. Rashin, A., Joffin, M., and Honig, B. (1986) *Biochemistry* **25**, 319–325.

17. Lewis, M., and Rees, D.C. (1985) *Science* **230**, 1163–1165.

18. Perutz, M.F. (1970) *Nature (London)* **228**, 726–739. Holbrook, J.J., Liljas, A., Steindel, S., and Rossman, M.G. (1975) *The Enzymes*, P.D. Boyer, ed. Academic Press, New York, pp. 191–292.

19. Bøje, L., and Hvidt, A. (1971) *Biopolymers* **11**, 2357–2364.

20. Warshel, A., Russell, S.T., and Churg, A.K. (1984) *Proc. Natl. Acad. Sci. U.S.A.* **81**, 4785–4789. Macgregor, R.B., and Weber, G. (1986) *Nature (London)* **319**, 70–73.

21. Rodgers, K., Pochapski, T., and Sligar, S. (1989) *Science* **240**, 1657–1659.

22. Lown, D.A., Thirsk, H.R., and Lord Wynne-Jones (1968) *Trans. Faraday. Soc.* **64**, 2073–2080. Neuman, R.C., Jr., Kauzmann, W., and Zipp, A. (1973) *J. Phys. Chem.* **77**, 2687–2691.

23. Xu, G.-J., and Weber, G. (1982) *Proc. Natl. Acad. Sci. U.S.A.* **79**, 5268–5271.

24. King, L., and Weber, G. (1986) *Biochemistry* **25**, 3632–3637.

25. Ruan, K., and Weber, G. (1989) *Biochemistry* **28**, 2144–2153.

26. Erijman, L., and Weber, G. (1991) *Biochemistry* **30**, 1595–1599. For previous observations of depolarization by energy transfer in protein adducts see Weber, G., and Daniel, E. (1966) *Biochemistry* **5**, 1900–1907. Weber, G., and Anderson, S.R. (1968) *Biochemistry* **8**, 361–371.

27. Weber, G., and Shinitzky, M. (1970) *Proc. Natl. Acad. Sci. U.S.A.* **65**, 823–830.

28. Hammes, G.G., and Kochavi, D. (1962) *J. Am. Chem. Soc.* **84**, 2069–2073. Hohnadel, D.C., and Cooper, C. (1972) *Eur. J. Biochem.* **31**, 180–185.

29. Weber, G. (1986) *Biochemistry* **25**, 3626–3631.

30. Lauberau, A., and Kaiser, W.A. (1975) *Annu. Rev. Phys. Chem.* **26**, 83–89. Heilweil, E.J., Casassa, M.P., Cavanagh, R.R., and Stephenson, J.C. (1989) *Annu. Rev. Phys. Chem.* **40**, 143–171.

31. Sturtevant, J.M. (1977) *Proc. Natl. Acad. Sci. U.S.A.* **74**, 2236–2240.

32. King, L., and Weber, G. (1986) *Biochemistry* **25**, 3637–3640. Dreyfus, G., Guimaraes-Motta, H., and Silva, J.L. (1988) *Biochemistry* **27**, 6704–6710.

33. Liddington, R., Derewenda, Z., Dodson, G., and Harris, D. (1988) *Nature* (*London*) **331**, 725–728.

34. Faber, H.R., and Matthews, B. W. (1990) *Nature* (*London*) **348**, 263–266.

35. Silva, J.L., Villas-Boas, M., Bonafe, C.S.F., and Meirelles, N.C. (1989) *J. Biol. Chem.* **264**, 15863–15867.

36. Gomes, F.C., Pereira, E.R., Bonafe, C.F.S., and Silva, J.L. (1990) In *Invertebrate Oxygen Carriers*, G. Preaux and L. Leontie eds. Leuven University Press, 315–318.

37. Anfinsen, C.B. (1967) *Science* **181**, 223–230.

38. Truesdell, C. (1985) *The Elements of Continuum Mechanics*, Parts IV and V. Springer-Verlag, Berlin.

39. Rogers, K.K., and Sligar, S.G. (1991) *J. Mol. Biol.* **221**, 1453–1460.

XVI

Biological Specificity and Ligand Binding

The complete description of molecular complexes involves three separate aspects: structure, energetics, and dynamics. An understanding of biological specificity necessitates a thorough knowledge of these three aspects to infer the relations between them. Structure, energetics, and dynamics are studied by very different methods and these areas have progressed to a large extent independently of each other. The first attempts at their synthesis are being carried out by means of lengthy, and up to now only tentative computer calculations; we shall give only a very brief account of their status after examining some significant experimental aspects that relate to the energetics and dynamics of the biological complexes.

Kinetic Significance of the Binding Energies

Table 1 shows a representative sample of the range of binding energies involved in complexes of proteins and small ligands. They are seen to vary between 20 and 3 kcal mol^{-1}, which correspond to dissociation constants from 10^{-15} to 10^{-2}.

In the simpler cases the dissociation constant is uniquely determined by the ratio of the rate-determining steps in the association and dissociation, k_+ and k_-, respectively. These quantities have been measured after initiation of the binding reaction by rapid mixing or by perturbation of an established equilibrium by flash photolysis and by changes in temperature, pressure, or electric field. From these observations a general rule about the relation of the thermodynamic equilibrium constant of the association reaction and the rates of association and dissociation that maintain the equilibrium may be derived. In the majority of cases the values of k_+ differ by only one to three orders of magnitude from the diffusion controlled rate, in other words there is no *large* enthalpy or entropy of activation linked to the association of the ligand. The relative large values of k_+ indicate that in the absence of the ligand the protein molecules

Table 1. Association Free Energies of Some Representative Complexes of Proteins and Ligands.

Protein	Ligand	DG	Source[a]
Avidin	Biotin	-20	Green (1963)
Lac repressor	DNA	-18	Riggs et al. (1970)
Antibody	Fluorescein	$-7, -14$	Voss (1984)
Horse liver ADH	NADH	-9	Theorell & McKinley-McKee (1961)
Beef heart LDH	NADH	-8.8	Anderson & Weber (1965)
Yeast ADH	Ethanol	-2.5	Mahler & Douglas (1957)
	Acetaldehyde	-4.7	
Fumarase	1-Malate	-2.5	Massey (1953)
	Fumarate	-3	

[a]References are given in [1].

have a conformation appropriate for its addition, in other words the ligand need not wait around for the gate to open, so to speak. The opposite would imply that protein molecules existed in a conformation inappropriate for binding and this less efficient device would be eliminated by selection in favor of molecules that accept the ligand almost equally well at all times, and therefore display rates of association that approach the diffusion controlled limit. Thus the 12 or more orders of magnitude in the change in dissociation constant displayed in Table 1 reflect primarily the different rates of dissociation of the ligand. Taking 10^7 mol^{-1} sec^{-1} as typical for k_+, k_- would vary in the examples of Table 1 from 10^{-5} sec^{-1}, one dissociation per month, to 10^6 sec^{-1}, or one dissociation per microsecond. The rates of dissociation of dimers and tetramer proteins at atmospheric pressure, estimated from the observed rates under higher pressures, are slow, of the order of 10^{-1} to 10^{-4} sec^{-1}, whereas the rates of association on decompression of pressure-dissociated adducts shows that k_+ cannot be much smaller than 10^6 mol^{-1} sec^{-1}. It appears that this characteristic relation between K and k_- is maintained in the case of the aggregation of monomers to form a dimer. When the subunits that form the oligomer are more than two the kinetic situation is more complex. We expect that the association involves the formation of intermediates, the simplest of which would be a dimer, which must be sufficiently stable to permit further addition of subunits for the association to proceed. The presence of successive association steps invalidates the usual assumption that the equilibrium results from two opposing processes characterized by unique rate constants. Furthermore a description of the equilibrium in these cases must take into account the time-dependent character of the processes involved, as we have repeatedly observed in the previous chapters.

In considering the very large differences in the rates of dissociation the conclusion is inescapable that these are determined by functional requirements: The role of avidin or of the antibodies is to reduce, respectively, the concentration of biotin, or that of the antigens to the lowest possible limits. Hemoglobin and myoglobin are very stable proteins in comparison with the unstable apoproteins, and the large free energy of association reduces the fraction of the unstable, and nonfunctional, apoprotein to a minimum. The free energies of formation of enzyme–substrate complexes are on the lower range, as the enzyme–product complex has to dissociate in very short time for efficient catalytic activity. In discussions of the absolute values of k_+ and k_- it is necessary to remember that the various spectroscopic methods by which these quantities are measured need not be equivalent and that conspicuously different values may be obtained when different methods are applied to the same system. For example, this will be the case if the association leads to a change in protein conformation and one of the methods used registers the association with the protein and another reveals the change in conformation that follows the binding. The existence of accommodation of the protein to the ligand as a result of binding can hardly be doubted both because of the small values of the elementary interactions within the protein and the experimental results on the protein dynamics (see below), but very few examples exist in which the changes that lead to such accommodation have been directly revealed by experiment.

Forces Responsible for the Specific Complexes

The many observations on the crystallography of protein–ligand complexes can be generalized in a single statement: The ligand is held by the protein through a number of interactions with almost all atoms of the ligand. In the cases that show the largest free energy of protein–ligand interaction (heme with apomyoglobin, or apohemoglobin, fluorescein with the specific antibodies, biliverdin with insectocyanin apoprotein, flavin in the various flavoproteins) the X-ray crystallography of the complex shows the ligand surrounded on nearly all sides by the protein and with only a minimal surface of contact with solvent [2]. Although a great many of the protein internal interactions result from simple van der Waals contacts, and very few involve permanent charges, the ligand–protein complexes with the higher free energies of interaction invariably show some with the latter characteristics. Close-range interaction over most of the ligand implies a complementary shape of binding site and ligand but short-range van der Waals contacts alone cannot confer a high degree of specificity to the interactions, as was originally thought following ideas first proposed by Pauling in 1948 [3]. The differences in the free energy of binding of NAD^+ and NADH by the dehydrogenases, or the strong binding of anions and much weaker binding of neutral molecules and cations by serum albumin are two among many examples that show the importance of ionic charges in both ligand and protein. The X-ray crystallographic data of protein–ligand complexes offer direct proof: Often a carboxyl or phosphate in the ligand is engaged at short range by an arginine in the protein and peptide carbonyls form hydrogen bonds with groups that would be recognized as proton donors in their hydrogen bonding to water. As these two types of bonds involve both saturation and direction they have a key role in determining the positioning and absolute orientation of the ligand with respect to the protein. It is also often found that the ligand in the binding site has one or more aromatic amino acid residues as neighbors and when the ligand is an aromatic molecule the orientation of the planes of the interacting rings are those that [4] make best use of the directional polarizabilities in the interaction.

Calculation of the total energy of the interaction in the protein ligand complex $(P - X)$ can be readily done from the relative positions of the atoms in the structure and a set of potential energy functions derived from consideration of the interactions of simpler molecules, but this set of interactions $(P - X)$ is insufficient to compute the free energy of association of protein and ligand: Previous to binding we have interactions between protein and solvent $(P - S)_0$, solvent and solvent $(S - S)_0$, and ligand and solvent $(X - S)_0$. After the ligand is bound all these quantities have changed into $(P - S)_1$, $(S - S)_1$, and $(X - S)_1$, respectively, and the changes have to be added to $P - X$. The change in energy, if not yet in free energy, when a molecule of the ligand is transferred from the solution to the binding site will be given by

$$dU = (P - X) + (P - S)_1 + (S - S)_1 + (X - S)_1$$

$$-\{(P - S)_0 + (S - S)_0 + (X - S)_0\} \tag{1}$$

Apart from the difficulties of determining each of the quantities in (1) that involve the solvent there is the additional difficulty of the precision required to derive a relatively small dU as the algebraic sum of the much larger values of the terms in Eq. (1). A better idea of the difficulties involved in an exact calculation may be obtained by considering a simple salt linkage involving a carboxyl and either arginine or lysine. From the fact that we are dealing with two fully ionized groups we might have concluded that the energy of their interactions is similar in both cases. However from the pKs of lysine and arginine we know that their free energies of ionization in water solution differ by over 4 kcal mol^{-1}. Thus the free energy change that follows the neutralization of charge from water at pH 7 or less will be much greater for arginine than for lysine. Consequently, the free energy change on formation of a salt linkage with arginine will be considerably greater than that corresponding to a salt bridge with lysine, and this difference may be the reason for the protein preference of one amino acid over the other in any given case. It should be noted also that the energy of the salt bridge in the protein will depend on the exact separation of the charged atoms. Equation (**2**) gives the change in energy of interaction dU resulting from small change in distance dr between the electronic charges

$$dU = e^2\left[r^{-1} - (r + dr)^{-1}\right] \tag{2}$$

With $e = 4.8 \times 10^{-10}$ esu, $r = 5$ Å, and $dr = 0.1$ Å we obtain $dU = 1.3$ kcal mol^{-1}. Thus, the most refined atomic coordinates provided by the X-ray data do not give us directly the means for calculating the exact energy of a salt bridge, or of some of the other atomic interactions within the protein. This insufficiency has been recognized in attempts to use the crystallographic data as the starting point for calculations of interaction energy [5].

In well-defined molecular complexes it is convenient, and to some extent justified by thermodynamic principles, to identify a particular structure as giving origin to a fixed chemical potential, but this task appears particularly difficult in the case of proteins. We have seen in chapters of this book many examples in which the differences in chemical potentials that introduce important differences in the physiological functions of proteins are on the order of 2–4 RT. The large number of degrees of freedom of the protein and the existence of internal equilibria in which the resultant is of much smaller magnitude than the opposing forces that maintain it ensure an almost permanent uncertainty in the attribution of fixed structural characteristics to the physiologically relevant energetic states.

Quite apart from the difficulties of performing exact, or even approximate, calculations of the molecular interactions involved in the specific complexes it must be acknowledged that the vagueness of the various contributions to binding is rather in the nature of the problem: Specific binding must result from the effects of many small contributions; one single large source of binding energy cannot be truly dominant if there is to be discrimination between close molecular species. Since no atom or group of atoms in the ligand or protein is all important it follows that molecules differing from the specific ligand by a single feature, such as replacement of a chemical group, will in many cases be bound as well as, and sometimes better than, the physiological ligand. The very origin of molecular recognition implies that it cannot be perfect in any sense, and this concept is usefully put to daily practice in enzymology and pharmacology. It may also have played a part in the evolution of binding systems: multiple small changes in the protein finally adjusting the binding energy to the optimum value for each particular case.

The apparent imperfection of the apparatus of molecular recognition in the proteins has even more important consequences: The amino acid replacements observed in proteins of different species that catalyze the same reaction can be very extensive without altering greatly the specific function. Without a doubt the most interesting fact to emerge from the studies of artificial amino acid replacements is the regularity with which negative results are obtained, even in many cases in which the replaced amino acid had been assigned indispensable mechanistic functions on the basis of previous estimates [6]. It is almost impossible to avoid the conclusion that proteins are systems of distributed functions in which no single amino acid plays a dominant role, except for those rare cases in which a chemical reaction results in a stable protein product. These observations clearly show that the reduction of protein function to the level of the isolated amino acids is not profitable and that we require a theory of heterogeneous systems that views functions as resulting from cooperative effects of many inputs rather than as a sum of fixed independent influences. Unfortunately such a task is not easy, as it runs contrary to the usual reductionist approach that asserts that the properties of the separate parts are all that is required to satisfactory explain the properties of an arbitrary system. Even if this premise is accepted in its largest and less precise meaning, it will do no harm to ask ourselves if complex systems cannot exhibit properties of matter that are not detected in experiments with the simpler systems that can be set up with the isolated components.

Binding Energies, Free Energy Couplings, and the Control of Metabolism

The first observers of biological specificity, enzymologists and immunologists, were particularly impressed by the discrimination in binding between closely related molecular species [7]. In view of the above remarks this property appears to us much less absolute, and from a biological point of view much less interesting, than the adjustment of the binding energy to a level that is optimal for the intended function. When the binding energy is small, and consequently the lifetime of the complex is short, it matters little whether some, usually small, fraction of the binding sites is filled by the wrong ligand; this shall soon leave and be replaced by the right one without rendering permanently ineffective the protein molecule to which it is attached. It is only when the binding energies exceed certain limits that attachment of the wrong ligand can have biologically undesirable consequences. We can look in the same way at the modulation of enzyme action by allosteric effectors: Let C_{phys} stand for the physiological concentration of a substrate and K for the dissociation constant of the Michaelis complex with a specific enzyme that transforms it. The useful range of regulation of the enzyme action is comprised between $C_{phys}/K = 0.1$ and $C_{phys}/K = 10$. For the lower concentration the enzyme works at 9% of capacity, and for the second at 90%. Between these limits regulation can be achieved by adjusting K to a new level as a result of free energy coupling between the binding of the substrate and the binding of an allosteric regulator, or by a change in C_{phys}. If the typical concentration is one that makes $C_{phys}/K = 1$ a free energy coupling of $+1.4$ kcal mol^{-1} with an allosteric regulator would reduce C_{phys}/K to 0.1. To restore function of the enzyme to the original level we require an increase in concentration of substrate of one order of magnitude. Had the free energy coupling been larger, e.g., 3 kcal, the substrate increase required for restoration of function would be over two orders of magnitude, and the enzyme could be considered as permanently shut off as long as the allosteric inhibitor was present. In consequence, a fine-tuned enzyme regulation that permits both the slowing down of the enzyme action by an allosteric effector and its restoration by a physiological increase in substrate concentration requires very moderate free energy couplings between substrate binding and allosteric modifier, in practice no bigger than 2 kcal mol^{-1}. Also the protein architecture must be such as to be able to transmit these small influences from one binding site to another. As in other cases of coupling of mechanical energy we face here the need for impedance matching with the transducing medium, the intervening protein structure. If the elementary bonds between the protein residues provided constraints capable of withstanding energies much larger than the free energy couplings the possibility of transmitting the influences from one binding site to the other would be lost. It follows that the small energies of interaction of the internal residues in proteins are indispensable for the efficient regulation of metabolism and that evolution has selected those protein structures that are appropriate for this purpose.

Dynamics of Protein Molecules

It is hardly possible to explain the mutual influence of ligands bound to a protein without some concomitant change in the protein structure. The small absolute magnitude of the free energy couplings between bound ligands predicates the possible appearance of the structures associated to varying degrees of coupling as a result of fluctuations of the thermal energy and of the volume of the protein [8]. The X-ray crystallographic analyses informs us of the average positions of atoms in the crystal, and in the best circumstances of the average amplitude of their thermal motions, but do not contain direct information as to the energy or frequency characteristic of those motions. The formation of the crystal itself may already modify the average solution structure or restrain some of the possible fluctuations. Although at present it is not possible to exactly assess the importance of such crystal forces they cannot be lightly dismissed for, after all, without them there would be no crystal. The observation that ligand binding of oxygen can cause shattering of crystals of deoxyhemoglobin when prepared according to one method [9] but not when prepared according to another [10] indicates the existence of variable reciprocal effects between inter- and intramolecular interactions. Various methods have been used to compare the properties of the molecules in the crystal and in solution. Presently, two-dimensional NMR [11] offers the most promising method for this purpose, as it seems possible to obtain by its means a completely independent estimate of a large number of distances between specific protons of the protein in solution. Until the recent appearance of the NMR methods, apart from a few observations of X-ray scattering of protein solutions [12], comparisons of predictions of solution behavior from the structural features of the crystals with actual experiments have been limited to a few favorable cases. The observation of a fast protein dynamics discussed below makes the assessment of the results more delicate and uncertain that appeared at the time when such protein dynamics was not taken into consideration. I shall mention only the observation that covalent conjugates of various forms of carboxypeptidase with the diazonium salt of arsanilic acid yield almost exclusively the tyrosin-248 azo-derivative [13]. Whereas the solutions are red on account of the formation of the Zn complex of the azo dye the crystals are yellow indicating that the protein conformations selected by crystallization prevent the interaction of the Zn atom with the azo chromophore. The further observation that similarly labeled carboxypeptidase can be prepared, which displays the red color of the Zn complex in both crystal and solution [14], stresses the fact that the protein conformations that favor or prevent the appearance of the red color cannot have very different energies and that external conditions may tip the balance one way or another.

Slow Protein Motions: Hydrogen Isotope Exchange

Direct kinetic methods have been used to investigate the time-dependent fluctuations of the protein molecules. The classical method is the isotope exchange of peptide bond hydrogens, originally proposed by Linderstrom-Lang and collaborators [15]. A prevalent interpretation of the earlier findings was that the appearance of highly disorganized forms of the protein as a result of a thermal fluctuation resulted in a rapid exchange of the peptide hydrogens of that fraction. This was deduced from the large energy of activation (> 35 kcal) computed for the process. More recent studies [16] of the effect of temperature on the isotope exchange have indicated that these large energies of activation for exchange are operative only if the range of temperatures employed in the measurements includes those (40–60°C) at which one expects occasional large thermal disruptions of the structure. A measure of the intrinsic thermal activation of the isotope exchange is obtained if this is followed over a smaller temperature range (5–25°C), at which overall reversible unfolding of the protein is a rare event. Though the final amount of isotope exchange is similar in these conditions, the rate of isotope exchange is slowed down, and the energy of activation turns out to be much smaller, under 10 kcal mol^{-1}. NMR has permitted the identification of individual protons [17] and the measurement of their rates of exchange in some of the smaller proteins. The vastly different rates observed indicate that proton exchanges are linked to local conditions rather than to a more general unfolding of the peptide chain. Does the exchange result from occasional local unfolding of segments of peptide chain that become exposed to solvent [18] or from the penetration of individual molecules of water into the actual protein structure [19]? In either case the exchange is a manifestation of the protein dynamics, but of very different physical origin and magnitude. To the present it has not been possible to devise experiments that clearly distinguish between these two possibilities and the precise dynamic features in the protein that facilitate the isotope exchange remains, in many cases obscure. On purely a priori grounds one may note that penetration of a molecule of water into the structure creates a much smaller additional solute–protein interface than exposure to solvent of even a small peptide region, and should be favored on such energetic grounds.

Fast Protein Motions: ^{13}C and Proton NMR, Fluorescence Spectroscopy, and Debye–Waller Factors

Other methods have been used to demonstrate the existence of a fast protein dynamics: Proton and carbon magnetic resonance on one hand and fluorescence spectroscopy on the other. High-resolution proton NMR was used originally by Wuthrich [20] to demonstrate the existence of slow and fast-rotating phenylalanines in proteins and the conversion of the former into the latter category as the temperature is increased. ^{13}C NMR has been used by Allerhand [21] to display the motions of the carbon skeleton of the protein in the nanosecond time domain. Fluorescence spectroscopy offers the simplest and most dramatic way to demonstrate the existence of protein dynamics in the nanosecond time range. Lakowicz and Weber [22] studied the quenching of protein fluorescence by oxygen. Oxygen is an ideal quencher of the singlet state, virtually every collision between fluorophore and oxygen resulting in quenching. The cross section for oxygen quenching appears to be that of the fluorophore itself so that direct interaction of oxygen and fluorophore, without interposition of solvent, is required for quenching. In eleven proteins studied quenching of tryptophan fluorescence by oxygen was found to be 10–20 times more effective than by iodide, although free tryptophan and its simple derivatives are quenched almost equally well by oxygen and iodide. This difference of the efficiency of the two quenchers of protein fluorescence all but disappears in 6 M guanidine hydrochloride solution. Thus, in the native proteins oxygen reaches the vicinity of the tryptophan residues with much higher efficiency than iodide, and to this purpose it must penetrate the native protein structure helped by its liposolubility and absence of electric charge. It was also uniformly found that the fluorescence lifetime and yield decrease proportionally, indicating that the quenching is a dynamic phenomenon taking place by the diffusive approach of oxygen to tryptophan during the fluorescence lifetime of the latter (2–6 nsec). The proteins with the bluer emissions, such as aldolase or azurin, in which the tryptophans are buried in the structure, are quenched almost as effectively as lysozyme or serum albumin, in which the redder-emitting tryptophans are exposed to the solvent. Penetration and diffusion of oxygen within the protein are not possible without fluctuations of the protein structure in the subnanosecond time scale. Separation of residues or atoms must take place to permit oxygen diffusion and the separated groups cannot be held together by energies larger than 2–4 kcal if they are to become temporarily separated by an energy fluctuation with a recurrence time of a fraction of a nanosecond. A more detailed analysis of the penetration and diffusion of oxygen within the structure aims at separating the rate of entry of oxygen in the protein from the actual rate of quenching by diffusion within the protein. From these figures and the absolute concentration of oxygen required for quenching one can estimate the partition coefficient of oxygen between protein and solvent. Jameson et al. [23] have carried out these calculations for the quenching of protoporphyrin fluorescence in the specific apomyoglobin adducts in which the heme group is replaced by the porphyrin (Des-Fe myoglobin). The diffusion of oxygen within hemoglobin or myoglobin can be deduced also from the existence of a well-known kinetic phenomenon occurring in the course of the recombination of oxygen or CO with the heme after a dissociating light flash [24]. It is found that a small fraction of the ligand dissociated by the flash recombines in the course of nanoseconds whereas the rest does so over times of microseconds and longer. The fast component, designated as geminate recombination, results from the diffusive rebinding to the heme of O_2 or CO molecules "stranded" within the protein structure after the flash. The figures obtained from the experiments on the quenching of the fluorescence of Des-Fe myoglobin by O_2 agree, as regards the magnitude of the rate of diffusion of O_2 within the protein, with those derived from the geminate recombination experiments.

As already discussed in Chapter XIII the motion of the tryptophan residues may be detected through observations of the depolarization of their fluorescence. Measurements of stationary fluorescence give a good indication of the amplitude of those motions that take place within the fluorescence lifetime, but real time measurements can go further and detect not only the amplitude and rate of these motions but the homogeneity of the emitting population as well. Differential polarized phase fluorometry has been used for this purpose [25] and the range of modulated excitation has been extended into the gigaherz region [26]. At 500 MHz a degree of phase shift of the fluorescence emission with respect to the excitation corresponds to a time delay of 5.5 psec. In proteins containing a single tryptophan Axelsen *et al.* [27] find that local rotations of some 20–30° of amplitude in lysozyme are accomplished in times of the order of 100 psec. Additionally the emission is heterogeneous as regards lifetime indicating the presence of two or more tryptophan environments that can exchange only within times longer than the rotational times.

The X-ray crystallographic data can also be analyzed to obtain an average of the amplitude of displacements of the atoms in the structure through the so-called Debye–Waller factors [28]. Evidently, only those displacements that occur with high frequency during the exposure of the sample to the X-rays could be thus detected. From these analyses it has been concluded that different regions in the protein vary considerably in their mobility. In myoglobin, for example [28], the atoms closest to the heme show the smallest displacements; other regions of the protein exhibit considerably higher mobility. This finding is not surprising as the heme provides the various amino acid residues with which it interacts with a point of attachment much more rigid than the protein's other amino acid residues.

The general conclusion from the results obtained by the different experimental methods is that the protein is the site of a lively *spontaneous* dynamics that extends into the nanosecond and even shorter time domains. From the experimental data of fluorescence spectroscopy one would judge that "detectable" motions of the structure do not occur in times shorter than 10–50 psec and that composite motions that disorganize the protein structure more radically must occur much more rarely. For the present we have no means of determining the effective cut-off high energy or low-frequency of such motions. It is doubtful that they present great physiological interest; the really important protein motions are *induced* motions that follow changes in the environment, particularly the increase in the concentration and consequent association of specific ligands.

The Numerical Simulation of Protein Dynamics

The advent of very fast, large capacity digital computers has stimulated the investigation of the possibility of deriving the specific architecture of proteins by piecing together the innumerable elementary interactions between the different parts of the peptide chain, and of following its dynamics through their changes in time under the influence of the thermal energy. For the reasons that we discussed in Chapter XIII the first of these aims seems still far from us, but the second, the simulation of the dynamics starting with the structure provided by the X-ray studies of proteins, has had a beginning of a realization. The method used for these simulations is that of "molecular dynamics" [29] in which the possible motions of every part of the molecule that is judged to move independently are determined from considerations of near neighbor interactions. These are governed by the potential functions giving the energy–distance relations for the various types of interactions (van der Waals, hydrogen bonds, electrostatics of permanent charges, torsional motions about bonds, etc.). The resultant of the interactions operating on each part of the molecule provides a trajectory direction that changes the initial coordinates. The permissible changes in coordinates are those that depend linearly on the time and this condition limits the time intervals during which a given set of forces is operative to a very small fraction of a picosecond. After this time interval the new coordinates of the moving parts are determined and new resultants for the forces are calculated.

Because of the very large number of cycles required for a trajectory covering a few picoseconds the method can be applied only to very short-term dynamics. In applications of this method to assemblies of free particles, as was done in the classical calculations of Rahman and Stillinger of liquid water (see Chapter X), a stationary condition (thermodynamic equilibrium?) can be reached in a short time and the equilibrium properties may be derived and compared with the experimental values. In proteins the problem is far more complicated: The number of total interactions is very much larger and the interactions are also of many types. One cannot start with any arbitrary distribution and expect to reach a final state of equilibrium distribution in the finite time, practically a few hours, in which the computer calculations have to be carried out. Additional limitations by the computer capacity as to the number of independently moving entities have restricted the investigation to models of proteins without hydrogen atoms surrounded by vacuum or by a continuum with interactive properties similar to water. The last restriction has been considered particularly unfavorable and a recent study surrounds the protein with a layer of 2500 molecules of water [30]. Explicit consideration of the hydrogens has been shown recently to lead to a better agreement between theoretical prediction and experimental results [27]. Some of the qualitative conclusions obtained by the molecular dynamics simulations are in agreement with both the experimental results and the conclusions derived from them: Thus the calculated *amplitude* of the rotational motion of tyrosine in bovine pancreatic trypsin inhibitor or tryptophan in lysozyme is of the same magnitude as the experimental ones. The possible diffusion of water molecules or of molecules of O_2 or CO within the structure of myoglobin as a result of structural fluctuations is also substantiated by the calculations. Finally the short-term displacements of the atoms calculated from the Debye–Waller thermal factors correlate reasonably well with those predicted by the short-term molecular dynamics.

On the other hand the purely dynamic information provided by molecular dynamics seems much less certain: The times for the local rotations of 20 or 30° of tyrosine and tryptophan are often predicted to be much faster (1 to 20 psec) than those experimentally observed (100 psec or more). The agreement with the Debye–Waller factors does not imply justification of the predicted time scale of the motions, as agreement would result from superposition of different conformers that exchange slowly in comparison with the very short times of the molecular dynamics simulation, yet the restrictions to motion that arise from interaction with the neighboring atoms would impose substantially the same limits in either case. It is of interest to recall that the fastest deactivations of simple vibrational states in liquids take place in 2 to 4 psec and that the more common figures range from 20 to 80 psec [31]. Evidently deactivation of these motions is not simply determined by encounters with less energetic molecules, and similar non-Newtonian restrictions are bound to be present in the interactions of the structural elements that make up the protein. In the many experimental methods by which the dynamics of molecules is studied the time appears as the ratio to a characteristic time and it is not clear to this author how an absolute scale of time valid in all circumstances can be attached to the elementary molecular and atomic motions.

A variety of experiments indicate that proteins can undergo slow changes in their properties in times that are evidently not accessed in the molecular dynamics simulations. However, the most basic assumption made in the application of molecular dynamics to proteins is that there are no restrictions to the changes in conformation of a protein other than those that arise from the most localized forms of atomic interaction. Following that premise we expect that all probable conformations of the protein are accessed in the times of the simulations, and that we cannot expect in proteins persistent heterogeneities unless they arise from permanent structural differences.

Taking all these circumstances into account it appears that molecular dynamics permits an estimation of the amplitudes allowed to the molecular motions and can thus be of help in steric and structural calculations. However, molecular dynamics does not seem able to reproduce the molecular events in the time scale in which they occur, even if we leave out of consideration the long-lived functional states that the study of oligomeric proteins has revealed. Molecular dynamics simulations do not immediately permit detection of what the upper limit of protein fluctuations can be, but it is noteworthy that even in the short time intervals explored the observed displacements have a distinctly periodic character, indicating that the truly harmonic motions of the protein are accomplished in very short times. The main contribution of molecular dynamics has been the demonstration that the elementary energetics and structure of proteins can be related—albeit in a crude way—to demonstrate the existence of a *spontaneous* subnanosecond dynamics in the protein molecules. It is this spontaneous dynamics that provides for the ability of the protein to undergo rapid changes in conformation following the interaction with ligands. In other words the spontaneous dynamics provides a basis for the physiologically all-important *induced* dynamics.

Biological specificity: Perspective and Prospects

The first interpretations of the specific affinities of proteins for small ligands and for other proteins relied on the "lock and key" analogy, which we can reinterpret in light of present knowledge as indicating the overwhelming importance of short-range interactions. The structural investigations of the complexes of nucleic acids and proteins have shown the complete success of this principle. However, we have to recognize that the formation of specific complexes is only the first step in the biologically important events that follow it, events that in one form or another involve the chemical properties of the substances, and that the subsequent chemistry has to be considered as a unity together with the association. We are certain that chemistry involves much more than simple steric factors and progress in this area of protein function will require a better knowledge of chemical reactivity than we possess today.

A "perfect" knowledge of proteins involves a unified explanation of the structure, energetics, and dynamics. It supposes that the macroscopic thermodynamic parameters, and the kinetic parameters that determine their physiological functions are all derivable from the application of fixed rules to the structural atomic characteristics of the protein. Our knowledge of the latter has advanced enormously in the past 50 years and has reached a stage that does not appear possible to improve, except in the direction of a higher precision of the atomic detail. On the other hand energetics and dynamics have remained practically at the phenomenological stage and it seems very doubtful that any increase in precision, or advances in methodologies, can lead us to an unequivocal identification of the structural substrates that are responsible for them. Future knowledge of the relation of protein function to structure and dynamics is much more likely to come from the comparative study of the proteins as integrated systems possessing certain biological and physical properties than from their study as isolated entities to which elementary physics and chemistry are applicable.

References

1. Green, N.M. (1963) *Biochem J.* **89**, 585–000. Riggs, A.D., Suzuki, H., and Burgeois, S. (1970) *J. Mol. Biol.* **48**, 67–83. Anderson, S.R., and Weber, G. (1965) *Biochemistry* **4**, 1948–1957. Herron, J.N., and Voss, E.W., Jr. (1984) In *Fluorescein Hapten: An Immunological Probe*, E.W. Voss, Jr., ed. CRC Press, Boca Raton, FL, pp. 77–98. Theorell, H., and McKinley-McKee, J.S. (1961) *Acta. Chem. Scand.* **15**, 1811–1833. Mahler, H.R., and Douglas, J. (1957) *J. Am. Chem. Soc.* **79**, 1159–1166. Massey, V. (1953) *Biochem. J.* **55**, 172–177. For some of the kinetic procedures see Kustin, K., ed. (1969) *Methods Enzymol.* **16**.

2. Padlan, E.A., Davies, D.R., Rudikoff, S., and Potter, E.M. (1976) *Immunochemistry* **13**, 945–949. Herron, J.N., He, X.-M., Masom, M.L., Voss, E.W., Jr., and Edmundson, A.B. (1989) *Proteins* **5**, 271–280. Bedzyk, W.D., Herron, J.N., Edmumdson, A.B., and Voss, E.W., Jr. (1990) *J. Biol. Chem.* **265**, 133–138.

3. Pauling, L. (1948) *Nature (London)* **161**, 707–709.

4. Herron, J.M. (1984) In *Fluorescein Hapten: An Immunological Probe*, E.W. Voss, Jr., ed. CRC Press, Boca Raton, FL, pp. 49–76.

5. Jensen, L.H. (1973) *Annu. Rev. Biophys. Bioeng.* **3**, 81–93.

6. Schimmel, P. (1990) *Biochemistry* **29**, 9495–9502. Alber, T. (1989) *Annu. Rev. Biochem.* **58**, 765–798.

7. Landsteiner, K. (1945) *The Specificity of Serological Reactions*, Chapter V. Harvard University Press, Cambridge, MA.

8. Cooper, A. (1976) *Proc. Natl. Acad. Sci. U.S.A.* **73**, 2740–2741.

9. Haurowitz, F. (1938) *Hoppe-Seyler Z. Physiol. Chem.* **254**, 266–274.

10. Liddington, R., Derewenda, Z., Dodson, G., and Harris, D. (1988) *Nature (London)* **331**, 725–728. Compare with Perutz, F.M. (1970) *Nature (London)* **228**, 726–739.

11. Wuthrich, K. (1989) *Science* **243**, 45–50 see also Wuthrich, K. (1989) *Acc. Chem. Res.* **22**, 36–44.

12. Stuehrman, H.B. (1973) *J. Mol. Biol.* **77**, 363–369. Doucet, J., and Benoit, J.P. (1987) *Nature (London)* **325**, 643–646.

13. Johansen, J.T., and Vallee, B.L. (1971) *Proc. Natl. Acad. Sci. U.S.A.* **68**, 2532–2535; (1973) *Proc. Natl. Acad. Sci. U.S.A.* **70**, 2006–2010.

14. Quiocho, F.A., McMurray, C.H., and Lipscomb, W.N. (1972) *Proc. Natl. Acad. Sci. U.S.A.* **69**, 2850–2854.

15. Hvidt, A., and Nielsen, S.O. (1966) *Adv. Protein. Chem.* **21**, 288–386. The original observation is that of Hvidt, A., and Linderstrøm-Lang, K. (1954) *Biochim. Biophys. Acta* **14**, 574–575; *Biochim. Biophys. Acta* **16**, 168–169.

16. Woodward, C., Simon, I., and Tuchsen, E. (1982) *Mol. Cell. Biochem.* **48**, 135–160. Wagner, G. (1983) *Quart. Rev. Biophys.* **16**, 1–87.

17. Wuthrich, K., Wagner, G., Richarz, R., and Braun, W. (1980) *Biophys. J.* **32**, 549–560.

18. Englander, S.W., Downer, N.W., and Teitelbaum, H. (1972) *Annu. Rev. Biochem.* **41**, 903–924.

19. Richards, F.M. (1979) *Carlsberg Res. Commun.* **44**, 47–53.

20. Wagner, G., Demarco, A., and Wuthrich, K. (1976) *Biophys. Struct. Mech.* **2**, 139–158.

21. Allerhand, A. (1978) *Acc. Chem. Res.* **11**, 469–474.

22. Lakowicz, J.R., and Weber, G. (1973) *Biochemistry* **12**, 4171–4179. Lakowicz, J.R., and Maliwal, B.P. (1983) *J. Biol. Chem.* **258**, 4794–4801.

23. Jameson, D.M., Gratton, E., Weber, G., and Alpert, B. (1984) *Biophys. J.* **45**, 795–803.

24. Alpert, B., El-Moslini, S., Lindquist, L.E., and Tfibel, F. (1979) *Chem-Phys. Lett.* **64**, 11–16. Lindqvist, L.E., El-Moslini, S., Tfibel, F., and Andre, J.C. (1981) *Chem. Phys. Lett.* **79**, 525–528. Henry, E.H., Sommer, J.H., Hofrichter, J., and Eaton, W.A. (1983) *J. Mol. Biol.* **166**, 433–451.

25. Jameson, D.M., Gratton, E., and Eccleston, J.F. (1987) *Biochemistry* **26**, 3894–3902.

26. Laczko, G., Gryczynski, I., Gryczynski, Z., Wiczk, W., Malak, H., and Lakowicz, J.R. (1990) *Rev. Sci. Instrum.* **61**, 2331–2337.

27. Axelsen, P.H., Gratton, E., and Prendergast, F.G. (1991) *Biochemistry* **30**, 1173–1179.

28. Frauenfelder, H., Petsko, G.A., and Tsernoglu, D. (1979) *Nature* **280**, 558–568.

29. Karplus, M. and McCammon, J.A. (1981) *CRC Crit. Rev. Biochem.* **9**, 293–349. Karplus, M. and Petsko, G.A. (1990) *Nature* **347**, 631–639.

30. Levitt, M. and Sharon, R. (1988) *Proc. Natl. Acad. Sci. U.S.A.* **85**, 7557–7561.

31. Lauberau, A. and Kaiser, W.A. (1975) *Annu. Rev. Phys. Chem.* **26**, 83–89. Heilweil, E.J., Casassa, M. P., Cavanaugh, R.R. and Stephenson, J.C. (1989) *Annu. Rev. Phys. Chem.* **40**, 143–171.

Index